Environmental Geomechanics
Problems and Sustainable Development

ABOUT THE BOOK

Global warming is one of the most serious threats that the planet Earth and its inhabitants may ever face. The Earth's climate has changed throughout the Earth's history—from glacial periods when ice covered significant portions of the Earth—to interglacial periods when ice either retreated to poles or melted entirely.

The planet is warming, from North Pole to South Pole, and everywhere in between. Globally, the mercury is already up more than 1 degree Fahrenheit (0.8 degree Celsius), and even more in sensitive polar regions. And the effects of rising temperatures aren't waiting for some far-flung future. They're happening right now. Signs are appearing all over, and some of them are surprising. The heat is not only melting glaciers and sea ice, it's also shifting precipitation patterns and setting animals on the move.

Although the science remains uncertain, 1990 marked the year when many governments accepted the impending reality of global climate change. Some governments began to announce targets for reductions in carbon dioxide in order to prompt the international negotiation process. Some governments remain unconvinced that the greenhouse effect is real. Others accept the science, but not the need to act dramatically. This volume is concerned not with the scientific debate as such, but with the basis for climate change policy in OECD countries. Crucial to the whole process of policy development is an assessment of the impacts of climate change.

ABOUT THE AOUTHOR

Naidoo Yaqub, finished her master in technology and currently pursing her PhD in environmental sciences from Institute of technology, Carlow, Irland. She is keen enthusiast of Environmental protection and Management. An alumunus of Indo-German centre for susutainibilty which is essential a collaborative interface to gather scientists in the feild of environmental sciences to come together and work for the comman cause which is environment protection.

Environmental Geomechanics
Problems and Sustainable Development

NAIDOO YAQUB

WESTBURY PUBLISHING LTD.
ENGLAND (UNITED KINGDOM)

Environmental Geomechanics Problems and Sustainable Development
Edited by: Naidoo Yaqub
ISBN: 978-1-913806-49-1 (Hardback)

© 2021 Westbury Publishing Ltd.

Published by **Westbury Publishing Ltd.**
Address: 6-7, St. John Street, Mansfield,
Nottinghamshire, England, NG18 1QH
United Kingdom
Email: - info@westburypublishing.com
Website: - www.westburypublishing.com

This book contains information obtained from authentic and highly regarded sources. All chapters are published with permission under the Creative Commons Attribution Share Alike License or equivalent. A Wide Variety of references are listed. Permissions and sources are indicated; for detailed attributions, please refer to the permission page. Reasonable efforts have been made to publish reliable data and information, but the authors, editors and publisher cannot assume any responsibility for the validity of the materials or the consequences of their use.

The publisher's policy is to use permanent paper from mills that operate a sustainable forestry policy. Furthermore, the publishers ensure that the text paper and cover boards used have met acceptable environmental accreditation standards.

Publisher Notice: - Presentations, Logos (the way they are written/ Presented), in this book are under the copyright of the publisher and hence, if copied/ resembled the copier will be prosecuted under the law.

British Library Cataloguing in Publication Data:
A catalogue record for this book is available from the British Library.

For more information regarding Westbury Publishing Ltd and its products, please visit the publisher's website- **www.westburypublishing.com**

Preface

Global warming is one of the most serious threats that the planet Earth and its inhabitants may ever face. The Earth's climate has changed throughout the Earth's history—from glacial periods when ice covered significant portions of the Earth—to interglacial periods when ice either retreated to poles or melted entirely.

The planet is warming, from North Pole to South Pole, and everywhere in between. Globally, the mercury is already up more than 1 degree Fahrenheit (0.8 degree Celsius), and even more in sensitive polar regions. And the effects of rising temperatures aren't waiting for some far-flung future. They're happening right now. Signs are appearing all over, and some of them are surprising. The heat is not only melting glaciers and sea ice, it's also shifting precipitation patterns and setting animals on the move.

Although the science remains uncertain, 1990 marked the year when many governments accepted the impending reality of global climate change. Some governments began to announce targets for reductions in carbon dioxide in order to prompt the international negotiation process. Some governments remain unconvinced that the greenhouse effect is real. Others accept the science, but not the need to act dramatically. This volume is concerned not with the scientific debate as such, but with the basis for climate change policy in OECD countries. Crucial to the whole process of policy development is an assessment of the impacts of climate change.

These impacts will appear first as physical changes in the environment, but will eventually translate into social and economic changes as well. Impacts are important, essentially because they represent the potential benefits that might be realised from taking political action against the climate change problem.

Global climate change has already had observable effects on the environment. Glaciers have shrunk, ice on rivers and lakes is breaking up earlier, plant and animal ranges have shifted and trees are flowering sooner.

This book is a highly informative source on global warming and climate change—issues that are perhaps the greatest threat to this planet. It will be a valuable reference tool for environmental scientists, activists and scholars in the field.

–Author

Contents

Preface (v)

1 Introduction   1

Global Warming: Causes And Resulting Climate Change/ Concentration of Greenhouse Gases in the Atmosphere / 20th Century Climate Change / 21st Century Climate Change / Atmospheric Aerosols / Agriculture / Evidence for Global Warming / Detection of Anthropogenic Global Warming / Attribution And The Fingerprint Method / Latitudinal Surface Temperatures / Tropospheric Warming and Stratospheric Cooling / Multivariate Fingerprints / When Will Attribution Occur / Future Climate Change / Gcm Climate Simulations / Greenhouse Feedbacks / Water Vapour Feedback / Cloud Feedback / Ice-Albedo Feedback / Greenhouse Gas Feedbacks / Greenhouse Modelling Versus Observation

2. Ozone Depletion   31

Introduction / What Causes Ozone Depletion? / How Long Has Ozone Depletion Been Occurring? / How Much of The Ozone Layer Has Been Depleted Around The World? / Ultraviolet Light and Ozone / Distribution of Ozone in the Stratosphere / Ozone Depletion / Why Is The Ozone Layer Important? / Oxygen And Ozone / Where is Ozone Found? / Other Effects of Ozone Depletion / Damage to Polymers / Effects on Biogeochemical Cycles / Effects on Climate / Montreal Protocol / The Protocol / Can Ozone Depletion Be Reserved / Phasing out Cfcs Are International Agreements Enough / Impact of Odss on Stratospheric Ozone / Impact of Temperature Changes on Ozone Chemistry / Impact of Methane and Water Vapour Changes on Ozone Chemistry / The Role of Transport for Ozone Changes / Stratosphere-Troposphere Dynamical Coupling / Possible Dynamical Feedbacks of Ozone Changes / Observed Changes in Stratospheric Ozone /

(viii)

Observed Changes in Odss / Observed Changes in Stratospheric Aerosols, Water Vapour, Methane and Nitrous Oxide / Observed Temperature Changes in the Stratosphere / Ozone in the Atmosphere and its Role in Climate / Chapter Outline / Ozone Depleting Substances / Hydrochlorofluorocarbons / Carbon Tetrachloride / Methyl Chloroform / Halons / Methyl Bromide / The Ozone Hole: Antarctica / Measuring the Ozone Hole / Why is the Hole Over The Antarctic / The Ozone Hole: Arctic / Do Ozone Holes Form Over the Arctic / Climate Change and Ozone Depletion in the Arctic / National Influences On Stratospheric Ozone / The Sun's Influence on Ozone / Atmospheric Winds and Ozone / Volcanic Eruptions and Ozone / Monitoring Ozone Depletion / Evidence for Stratospheric Ozone Depletion / Monitoring of Ozone Depletion / Current Understanding of Past Ozone Changes / Mid-Latitude Ozone Depletion / Winter-Spring Polar Depletion / The Montreal Protocol, Future Ozone Changes and Their Links to Climate

3. **Climate Change** 86

Terminology / Causes / Plate Tectonics / Solar Output / Orbital Variations / Volcanism / Ocean Variability / Human Influences / Astronomical Theory and Abrupt Climate Changes / Solar Variability and The Total Solar Irradiance / Biogeochemistry and Radiative Forcing / Climate Change Mitigation / Climate Change and Ecosystems / Forests / Mountains / Ecological Productivity / Time Scale of Climatic Change / C l i m a t e Sensitivity / The Effects of Climate Change on Ecosystems / Impact of Climate Change on Agriculture / Physical Evidence for Climatic Change / Historical and Archaeological Evidence / Glaciers / Vegetation / Ice Cores / Dendroclimatology / Pollen Analysis / Insects / Sea Level Change / Measurement of Climate Elements / Measurement of Temperature / Measurement of Rainfall / Measurement of Humidity / Measurement of Wind / Homogeneity / Statistical Analysis of Instrumental Records / Palaeoclimate Reconstruction From Proxy Data / Historical Records / Ice Cores / Stable Isotope Analysis / Physical and Chemical Characteristics of Ice Cores / Dating Ice Cores / Dendroclimatology / Ocean Sediments / Changes in Glaciers, Ice Sheets, and Ice Shelves / Modern Glacier Retreat / Antarctic Ice Shelf Breakups / Records of Climate Change / Radiative Forcing and Climate Sensitivity / Direct Radiative Forcing of Odss and Their Substitutes / Indirect Radiative Forcing of Odss / Net Radiative Forcing

4. **Effects Of Ozone Depletion And Global Warming**   134

   Ozone Depletion / Quantitative Understanding of the Chemical Ozone Loss Process / Public Policy In Response to the Ozone Hole / Current Events And Future Prospects of Ozone Depletion / The Rowland-Molina Hypothesis / the Ozone Hole / Ozone Depletion and Global Warming / Misconceptions About Ozone Depletion / Cfcs Are "Too Heavy" to Reach the Stratosphere / Man-Made Chlorine is Insignificant Compared to Natural Sources / An Ozone Hole Was First Observed in 1956 / if the Theory Were Correct, The Ozone Hole Should Be Above the Sources of CFCS / Consequences of Ozone Layer Depletion Increased UV / Biological Effects of Increased UV and Microwave Radiation From A Depleted Ozone Layer / Effects of Ozone Layer Depletion on Humans / Effects on Crops / Effects on Plankton / Observations on Ozone Layer Depletion / Chemicals in the Atmosphere / Verification of Observations / the Ozone Hole and its Causes / Interest in Ozone Layer Depletion / Effects on Earth's Food Chain / Action Plan for Organizations Using Refrigerants / Modelling Global Warming / Nitrous Oxide / Rainfall / Sea Level / Temperature / Trees / Water / the Truth About Global Warming / The Answer is Simple- Global Warming / Control the Harm Already Done

5. **Effects of Global Warming on Ecosystems**   166

   Results of Global Warming on Ecosystems / Observed and Expected Effects On Ecosystems / Polar and Ice-Related Changes / Fire And Drought / Biological Changes / Physical Changes Ecosystem Services in Global Warming / Ecology / Economics / Management and Policy / Vertical Forcing Patterns and Surface Energy Balance Changes / Spatial Patterns of Radiative Forcing / Alternative Methods of Calculating Radiative Forcing / Linearity of The Forcing-Response Relationship / Efficacy and Effective Radiative Forcing / Efficacy and the Forcing-Response Relationship / Basic Components of the Ecosystem / Incomplete Ecosystems / Global Warming Stress to Coral Reefs / The Effects of Global Warming on Open Oceans / Seaweed — A Carbon Sink / The Effects of Global Warming on Coastal Locations / Temperature

Index   205

Chapter 1

Introduction

Measurements of temperature taken by instruments all over the world, on land and at sea have revealed that during the 20th century the Earth's surface and lowest part of the atmosphere warmed up on average by about 0.6°C. During this period, man-made emissions of greenhouse gases, including carbon dioxide, methane and nitrous oxide have increased, largely as a result of the burning of fossil fuels for energy and transportation, and land use changes including deforestation for agriculture. In the last 20 years, concern has grown that these two phenomena are, at least in part, associated with each other. That is to say, global warming is now considered most probably to be due to the increases in greenhouse gas emissions and concurrent increases in atmospheric greenhouse gas concentrations, which have enhanced the Earth's natural greenhouse effect. Whilst other natural causes of climate change can cause global climate to change over similar periods of time, computer models demonstrate that in all probability there is a real discernible human influence on the global climate.

If the climate changes as current computer models have projected, global average surface temperature could be anywhere from 1.4 to 5.8°C higher by the end of the 21st century than in 1990. To put this temperature change into context, the increase in global average surface temperature which brought the Earth out of the last major ice age 14,000 years ago was of the order of 4 to 5°C. Such a rapid change in climate will probably be too great to allow many ecosystems to suitably adapt, and the rate of species extinction will most likely increase. In addition to impacts on wildlife and species biodiversity, human agriculture, forestry, water resources and health will all be affected. Such impacts will be related to changes in precipitation (rainfall and snowfall), sea level, and the frequency and intensity of extreme weather events, resulting from global warming. It is expected that the societies currently experiencing existing social, economic and climatic stresses will be both worst affected and least able to adapt. These will include many in the developing world, low-lying islands and coastal regions, and the urban poor. The Framework Convention on Climate Change and the Kyoto Protocol represent the first steps taken by the international community to protect the Earth's climate from dangerous man-made interference. Currently, nations have agreed to reduce greenhouse gas emissions by an average of about 5 per cent from 1990 levels by the period 2008 to 2012. The UK, through its Climate Change Programme, has committed itself to a 12.5 per cent cut in greenhouse gas emissions. Additional commitments for further greenhouse gas emission reduction will need to be negotiated during the early part of the 21st century, if levels of greenhouse gas concentrations in the atmosphere are to be stabilised at reasonable levels. Existing and future targets can be achieved by embracing the concept of sustainable development - development today that does not compromise the development needs of future generations. In practical terms, this means using resources, particularly fossil-fuel-derived energy, more efficiently, re-using and recycling products where possible, and developing renewable forms of energy which are inexhaustible and do not pollute the atmosphere.

Global warming begins when sunlight reaches Earth. The clouds, atmospheric particles, reflective ground surfaces and ocean surface then reflected about 30 per cent of it back into space, while the remaining is absorbed by oceans, lands and air. This in turn heats the planet's surface and atmosphere, making life possible. As Earth warmed up, this solar energy is radiated by thermal radiation or infrared heat, travelling directly out to space, thus cooling the Earth. However, some of the outgoing radiation is re-absorbed by carbon dioxide, water vapour and other gases in the atmosphere and is radiated back to Earth's surface; these gases are known as greenhouse gases due to their heat-trapping capacity. This re-absorption process is naturally good; the Earth's average surface temperature would be very cold if not for the greenhouse gases.

GLOBAL WARMING: CAUSES AND RESULTING CLIMATE CHANGE

The most authoritative reports on the causes and consequences of climate change come from the IPCC, particularly its 1995 Second Assessment Report (SAR) and its 2001 Third Assessment Report (TAR). The latter report refined the findings of the first assessment, pointing out that climate change is likely to be worse and occur more rapidly than initially predicted. Here I summarize the IPCC's findings on global warming and the worldwide effects of climate change before pointing out some of the anticipated socio-economic impacts in East Asia.

Introduction

According to the IPCC's TAR, there is now a collective picture, derived from an increasing body of observations, of a warming world and other changes in the Earth's climate system. The global average surface temperature increased during the twentieth century, with the 1990s and early 2000s the warmest on record. Snow and ice cover have decreased, global average sea level has risen, and the heat content of the oceans has increased.

Other aspects of climate have changed during the twentieth century, including changes in precipitation (e.g. increased heavy precipitation events) and cloud cover; fewer extreme low-temperature periods and more high-temperature periods; more frequent, persistent, and intense episodes of the El Nino ocean-warming event (and related adverse effects on weather in many areas); and an increase in areas experiencing drought and severe wet periods.

Some climate related events, such as tornadoes or tropical storms, do not appear to have changed based on IPCC data, although the evidence is conflicting. The TAR also finds that emissions of GHGs from human activities are altering the atmosphere in ways that are expected to affect climate.

Human activities have increased atmospheric concentrations of GHGs (e.g. CO_2, methane, nitrous oxide, halocarbons) and their warming potential. According to the report, "atmospheric concentration of carbon dioxide (CO_2) has increased 31 percent since 1750. The present CO_2 concentration has not been exceeded during the past 420,000 years and likely not during the past 20 million years. The current rate of increase is unprecedented during at least the past 20,000 years".

Three-quarters of human-induced emissions of CO_2 over the last two decades has come from the burning of fossil fuels (e.g. coal, oil, and natural gas), with most of the remainder the consequence of land-use changes, particularly deforestation. Natural causes of climate change have been relatively small. Furthermore, models for predicting future climate are increasingly accurate and precise. While uncertainties remain, understanding of climate processes and predicted effects has improved.

According to the TAR, new and stronger evidence points to human activities as the sources of observed global warming over the last fifty years, further strengthening the SAR's conclusion that the "balance of evidence suggests a discernible human influence on global climate". Warming over the last 100 years is unlikely to have been natural, with studies showing that global warming, particularly during the last 35-50 years, most likely resulted from human activities.

Thus, the TAR concludes: "In light of new evidence and taking into account the remaining uncertainties, most observed warming over the last fifty years is likely to have been due to the increase in GHG concentrations. Furthermore, it is very likely that the 20th century warming has contributed significantly to the observed sea level rise ... and widespread loss of land ice". Furthermore, the TAR determined that human activities will continue to shape the Earth's atmosphere throughout this century and into the future, and average global temperatures and sea levels are projected to rise. Emissions from burning fossil fuels will be the dominant source of atmospheric CO_2 during this century. These emissions and those of other GHGs would have to be reduced to "a very small fraction of current emissions" to stabilize climate.

Global average temperature is projected by the IPCC to increase by 1.4-5.8 degrees Celsius during this century (more than anticipated in the SAR). This warming will occur at a rate faster than that observed in the twentieth century, "very likely to be without precedent during at least the last 10,000 years". During this century, warming is expected to occur in most areas, but it should be particularly pronounced at northern high latitudes during winter.

Global mean sea level is expected to rise 0.09-0.88 metres in this century, with other very likely changes to include higher maximum temperatures and more hot days over most land areas, higher minimum temperatures and fewer cold days over most land areas, more intense precipitation events over many areas, increased summertime continental drying and drought over mid- latitude continental interiors, and more severe storms over some areas.

Ecological and Socio-economic Impacts of Climate Change

The ecological and socio-economic impacts of climate change are likely to be very significant and often painful. The TAR's findings on these impacts include the following: Regional changes in climate have already affected many physical and biological systems, with temperature increases being the most proximate cause.

Observed changes in regional climate have occurred in terrestrial, aquatic, and marine environments, and effects have included shrinking glaciers, thawing permafrost, reduced periods in which lakes and rivers are frozen, longer mid- and high-latitude growing seasons, shifts in animal and plant ranges to higher latitudes and altitudes, declines in populations of some animals and plants and reduced egg-laying in some birds, and insects populating new areas.

It appears that some social and economic systems have already been affected by increased floods and drought, but separating these ecological events from socio-economic factors is difficult.

Introduction

The TAR shows that many human systems are sensitive to climate change, including water resources, agriculture, coastal zones and marine fisheries, settlements, energy, industry, financial services (e.g. insurance industries affected by increased claims), and human health.

Adverse impacts of climate change include reduced crop yields in most tropical and sub-tropical regions; decreased water availability in many water-scarce areas, especially the sub-tropics; more people exposed to increased increase in flood risk from rising sea levels; and increasing demand for energy to cool areas affected by higher summer temperatures.

Some impacts may be positive, such as increased crop yields in some mid-latitude areas; potentially more timber if forests are managed appropriately (although increased pests could more than offset this); increased water availability for some water scarce areas; lower winter mortality in traditionally cold areas; and reduced winter demand for energy due to higher winter temperatures.

Many of the risks are unclear, and there is substantial potential for "large-scale and possibly irreversible impacts" from changing ocean currents, melting ice sheets, accelerated global warming due to atmospheric feedback effects, and so forth. In addition to efforts to mitigate climate change, the TAR argues that adaptation is a necessary strategy.

However, those people and societies with the least resources are most vulnerable because they are least able to adapt. Projected warming may result in a mixture of economic gains and losses for developed countries, but developing countries can expect mostly losses: "The projected distribution of economic impacts is such that it would increase the disparity in well-being between developed countries and developing countries," with the disparities increasing the greater the temperature increases. The upshot is that "More people are projected to be harmed than benefited by climate change," even if temperature increases are limited.

The TAR is not restricted to scientific and economic assessments. It argues that international justice and equity are important considerations when addressing climate change: "Inclusion of climatic risks in the design and implementation of national and international development initiatives can promote equity and development that is more sustainable and that reduces vulnerability to climate change".

Global Warming and Climate Change Impacts in East Asia

Clearly, the global effects of climate change are potentially major, and will likely lead to many adverse consequences, difficult choices, and expensive adaptation measures for much of the world's population. The countries of

East Asia will not be immune to these changes, and in most cases will be among the worst affected due to their vulnerable geographies and economies.

Effects may not always be adverse, but even if they are not they will likely increase unpredictability and require adaptation. What are the expected impacts of climate change in East Asia? Several research reports have anticipated the effects of climate change for the region. Some of their findings are summarized here to convey the scale and nature of the potential changes.

According to a 1997 report from the IPCC on anticipated regional impacts of climate change, temperate Asia (including Japan, the Koreas, and most of China) has experienced an average annual temperature increase of more than 1 degree Celsius in the last century, mostly since the 1970s, with substantial warming expected in this century. Rainfall is expected to change in the area, with substantial declines expected in most of China (notably northern provinces). Permafrost in northeast China is expected to disappear (with release of methane, thus adding GHGs to the atmosphere) and glaciers will melt. Northern China is particularly vulnerable to expected changes in rainfall, exacerbating existing water shortages.

The area is likely to experience changing agricultural yields, with many crops likely to see reductions and a northward movement of crop zones and anticipated shortages of roundwood (partly due to increased demand). Delta coastlines in China "face severe problems" from sea-level rise, which will include salt water intrusion into aquifers. Japan will not be immune; already many parts of major coastal urban areas, with millions of residents, are below the mean high-water mark.

Providing protection for only some of these cities will cost tens of billions of dollars. Japan's beaches, which comprise about a quarter of its coastline, will be subject to erosion - and over half of existing beaches may disappear. Additionally, heat-related deaths throughout temperate Asia may increase sevenfold by the middle of this century.

The potential effects of climate change for tropical Asia (encompassing Southeast Asia) are also described in the IPCC's 1997 regional report. It points out that the region already suffers from increasing pollution, land degradation, and all manner of environmental problems resulting from rapid urbanization, industrialization, and economic development. Climate change will exacerbate these problems. In this area, mean temperatures have already gone up by 0.3- 0.8 degrees Celsius over the last 100 years. Forest cover will change as a consequence, possibly increasing, and forest types may change. Changes in evaporation and rainfall are likely to have detrimental effects on freshwater wetlands. Coastal areas will be most greatly affected by sea-level rise and increased ocean temperatures (the latter possibly preventing coral reefs from keeping up with sea-level rise). Mangrove and tidal wetlands will have difficulty adapting due to bordering infrastructure and human activities. Greater erosion, coastal flooding, and salinization of fresh water sources are probable. Delta regions of Southeast Asian countries are particularly vulnerable, and throughout this area several million people could be displaced by sea-level rise.

The costs of responding to the impacts of rising seas, in the words of the IPCC, could be immense. Glaciers feeding the area's rivers will melt, and there may be yearly reductions - albeit between seasonal flooding - in the flow of snow-fed rivers, adversely affecting agriculture, hydropower generation, and urban water supplies.

temperature and moisture changes and possibly from increased pests, affecting, for example, wheat, rice, and sorghum crops (although much uncertainty, confounding planning, will obtain). According to the IPCC report, poor rural populations depending on traditional forms of agriculture or living on marginal lands are especially vulnerable. Increased vector-borne diseases such as dengue, malaria, and schistosomiasis will adversely affect human health in this area.

A 1999 report on climate change impacts prepared by Britain's Climatic Research Unit summarized many potential impacts for some of the countries of East Asia. In China, temperature increases are predicted to be greatest over northern areas, with changes in precipitation and threats to biodiversity. In Indonesia, forest fires are predicted to increase and endangered species may be threatened.

In Japan, heat waves will increase in frequency and intensity, coastlines and coastal infrastructure will be harmed, and reefs will be stressed. In the Philippines, rainfall will increase during the wet season and decrease during the dry season, reefs will suffer from warming water, and potentially millions of people will be threatened by sea-level rise.

Von Hippel has summarized a few of the possible impacts of climate change in Northeast Asia: pressure on agricultural resources and accelerated desertification leading to cross-border migrations, particularly from China to Russia; adverse climatic effects on North Korea's food production, possibly increasing military pressure on South Korea and creating economic burdens for reunification; increased demand for air conditioning, leading to higher fuel consumption and hence more local and regional air pollution (and adding still further to GHG emissions); salinization of breeding grounds for fish from sea-level rise, leading to reduced fishery yields that could exacerbate conflicts over marine resources; additional oil pollution (from shipments of oil imports) that may strain relations among countries sharing marine resources and shipping lanes; and increased economic costs from natural disasters like catastrophic storms, straining emergency and disaster relief resources in the region.

The 2001 TAR assessment of vulnerability in Asia shows that the region is potentially more susceptible to climate change than are some other regions of the world. It concludes that the developing countries of Asia are highly vulnerable to climate change, and their adaptability is low. (Developed countries of the region (e.g. Japan) are of course less vulnerable because they are more able to adapt to climate change.)

Floods, forest fires, cyclones, droughts and other extreme events have increased in temperate and tropical Asia. The TAR anticipates that while agricultural productivity could increase in northern parts of Asia, food security would suffer in arid, tropical, and temperate Asia due to reduced agricultural and aquaculture productivity from warmer water, sea-level rise, floods, droughts, and cyclones.

Water availability may decrease in arid and semi-arid Asia and possibly increase in northern Asia, and increased incidence of vector-borne diseases and heat-stress will threaten human health. Temperate and tropical Asia should anticipate increased rainfall and floods, and sea-level rise and more intense storms could "displace tens of millions of people in low-lying coastal areas of temperate and tropical Asia". Some parts of Asia will see climate change effects on transport, increased demand for energy, and adverse impacts on tourism. Land-use and land-cover changes will threaten biodiversity, and sea-level rise will adversely impact coral reefs and mangrove areas that are important for fisheries.

What comes from these (and other) reports on the impacts of climate change in East Asia is that many of the effects will be felt most by - and be most painful for - the developing countries of the region. They are generally more vulnerable and least able to cope due to poverty and existing environmental problems and resource scarcities.

A very large number of people throughout East Asia live in low-lying coastal regions, and they are threatened by sea-level rise, land subsidence, inundation of fresh water aquifers by salt water, and more frequent and violent storms from climate change. Island countries such as Indonesia and the Philippines are especially vulnerable to climate change effects.

They can expect freshwater shortages and damage to coastal areas and adjacent infrastructure, with concomitant adverse effects on tourism. (Indeed, in extreme cases it may one day be necessary for some small-island states to abandon their territory altogether. Representatives from these countries have for some time argued that they are *already* feeling the effects of rising oceans.

Introduction

Other poorer countries in the region are vulnerable. For example, the World Bank reported that Chinese research has estimated that a 1-metre rise in sea level would inundate 92,000 square kilometres of China's coast, displacing 67 million people (and more as population increases). According to one assessment, future climate change will reduce soil moisture in China, particularly in the north, and this will increase the demand for agricultural irrigation, which will in turn add to existing severe water shortages.

In short, "Possible impacts of climate change on Chinese agriculture could be highly disruptive ...". Already vulnerable, China may also see greater weather extremes, including droughts in the north and floods in the south, and heat stroke and death will increase, as may occurrences of malaria, dengue fever, and other diseases. An ever-growing body of research shows that climate change will (and probably has already) adversely affected human health, and this is particularly true of East Asia. Southeast Asia is especially vulnerable to anticipated increasing incidence of vector-born diseases. Hotter weather will increase heat-related mortality in the region, as indicated by historical studies from China showing a strong correlation between peak summer temperatures and death rates.

But even the developed countries and regions of East Asia are unlikely to avoid harm from climate change. For example, while Japan's coastlines are not as vulnerable as those of China, the Philippines, and other countries, it is reasonable to expect that it will suffer costly damage from sea-level rise, associated storm surges, and adverse weather, and it has direct interests in the health of surrounding seas and indirect interest in what happens throughout the region.

By way of example, recently reduced fish catches by Japanese fisherman have been attributed to changes in underwater currents triggered by global warming. And there will be adverse impacts for Japan's biodiversity, forests, agriculture, wetlands, and water systems, as well as for infrastructure and human health. According to the government, climate change effects have already become visible in Japan. For these and other reasons, Japan supports the climate change regime and the Kyoto Protocol- despite its greater ability to cope with climate change compared to its neighbours.

CONCENTRATION OF GREENHOUSE GASES IN THE ATMOSPHERE

The problem begins when the concentration of greenhouse gases in the atmosphere were artificially raised by humankind at an ever-increasing rate since the past 250 years. As of 2004, over 8 billion tons of carbon dioxide was pumped out per year; natural carbon sinks such as forests and the ocean absorbed some of this, while the rest accumulated in the atmosphere. Millions of pounds of methane are produced in landfills and agricultural decomposition of biomass and animal manure. Nitrous oxide is released into the atmosphere by nitrogen-based fertilizers and other soil management practices. Once released, these greenhouse gases stay in the atmosphere for decades or longer. According to the Intergovernmental Panel on Climate Change (IPCC), carbon dioxide and methane levels have increased by 35 and 148 per cent since the

1750 industrial revolution. Paleoclimate readings taken from ice cores and fossil records dating back to 650 000 years show that both gases are at their highest levels. Thermal radiation is obstructed further by the increased concentrations of greenhouse gases, resulting in what is known as enhanced global warming.

Recent observations of global warming have solidified the theory that it is indeed an enhanced greenhouse effect that is causing the world to warm. The planet has experienced the largest increase in surface temperature over the last century. Between 1906 and 2006, the Earth's average surface temperature rose between 0.6 to 0.9 degrees Celsius; the last 50 years saw the temperature increase rate almost doubling. Sea levels have shown a rise of about 0.17 metres during the twentieth century. The extent of Arctic sea ice has steadily shrunk by 2.7 per cent per decade since 1978, just as world's glaciers steadily receded.

As the world continues to consume ever more fossil fuel energy, greenhouse gas concentrations will continue to rise, and with them Earth's temperature. The IPCC estimates that based on plausible emission scenarios, average surface temperatures could increase between 2°C and 6°C by the end of the 21st century. Continued warming at current rates poses serious consequences. Low-lying coastal regions, with dense population, are especially vulnerable to climate shifts, with the poorer countries and small island nations having the hardest time adapting. It has been projected that by 2080, 13 to 88 million people around the world would lose their home to floods.

20TH CENTURY CLIMATE CHANGE

There is now considerable evidence that indicates a relationship between the man-made enhancement of the Earth's natural greenhouse effect through greenhouse gas pollution of the atmosphere and the global warming that has been observed.

Measurements of surface temperature recorded around the world during the last 150 years indicate that global temperatures are now higher than in any decade over this period. During the 20th century a global average surface temperature increase of about 0.6°C years has taken place, although the

Introduction

warming trend has not been smooth and has taken place rather differently between the Northern and Southern Hemispheres. With this in mind, it is important to recognise that global average surface temperature, as a measure of the global climate, represents an over-simplification. Winter temperatures and night-time minimums for example, may have risen more than summer temperatures and daytime maximums.

With higher global temperatures one would expect an increase in rainfall and other forms of precipitation, because of the greater amount of moisture available within the atmosphere. Striking changes in precipitation have occurred on regional scales, most notably the drought in the African Sahel between the 1960s and 1980s. Nevertheless, the accuracy of many precipitation records should be treated with caution. Precipitation is more difficult to monitor than temperature due to its greater geographical variability.

Variations in land- and sea-ice coverage and the melting or growth of glaciers occur in response to changes in temperature, sunshine, precipitation, and for sea-ice changes in wind. Since 1966 Northern Hemisphere snow cover maps have been produced by the United States using satellite imagery. Consistent with the surface and tropospheric temperature measurements is the decrease (by about 10 per cent) in snow cover and extent since the late 1960s. There has been a widespread retreat of mountain glaciers in non-polar regions during the 20th century. Variations in sea-ice extent have also been reported, with spring and summer sea-ice extent in the Northern Hemisphere decreasing by between 10 an 15 per cent since the 1950s. Considerable interest is now focusing on Antarctica, where regional warming, as predicted by climate models, has been more rapid than global warming as a whole. In recent years the summertime disintegration of the Larsen Sea Ice Shelf adjacent the Antarctic continent has been occurring on an unprecedented scale. In view of the rapidity at which it is taking place, such an event has been viewed as a possible signal of global warming.

Increased global cloudiness, as for increased global evaporation and precipitation, would be an expected consequence of higher global temperatures. It is likely that there has been a 2 per cent increase in cloud cover over mid - to high latitude land areas during the 20th century. In most areas the trends relate well to the observed decrease in daily temperature range (since cloudier nights tend to be warmer and cloudier days cooler).

21ST CENTURY CLIMATE CHANGE

Prediction of climate change over the next 100 to 150 years is based solely on climate model simulations run on computers. The vast majority of modelling has concentrated on the effects of continued man-made pollution of the atmosphere by greenhouse gases, and to a lesser extent, atmospheric aerosols. The main concern at present is to determine how much the Earth will warm in the near future.

Significant results from some of the best climate models available indicate that a global average warming of 0.3°C per decade can be expected to occur during the 21st century, assuming that mankind fails to control current emissions of greenhouse gases, although it could be as high as 0.6°C. In addition regional variations in the patterns of temperature and precipitation change will occur, with greater warming likely in the polar regions. Currently, models suggest that if the atmospheric concentration of carbon dioxide, the main greenhouse gas, is doubled from pre-industrial levels, the Earth will warm by between 1.5 and 4.5°C sometime over the next 200 years or so. The large margin of error in future prediction of temperature emphasises that modelling the climate is inherently a difficult business. Part of the problem stems from trying to guess what climate feedbacks might occur that may enhance the initial warming due to an enhanced greenhouse effect. Melting ice in the polar regions for example could accelerate warming because exposed ground absorbs more energy from the Sun than snow and ice, which reflect about 80 to 90 per cent.

Whilst uncertainties concerning the actual response of the global climate to man-made greenhouse gas emissions exist, most scientists agree that the global warming trend of the 20th century will continue into the 21st century. The projected rate of warming is faster than at any time during recent Earth history. If nations fail to respond, the world may experience numerous adverse impacts as a result of global warming in the decades ahead.

ATMOSPHERIC AEROSOLS

Atmospheric aerosols are very fine particles suspended in air. They are formed by the dispersal of material at the Earth's surface (primary aerosols), or by reaction of gases in the atmosphere (secondary aerosols). They include sulphates and nitrates from the oxidation respectively of sulphur dioxide and nitric oxide during the burning of fossil fuels, organic materials from the oxidation of volatile organic compounds (VOCs), soot from fires, and mineral dust from wind-blown processes. Natural aerosols, which also include sea salt and volcanic dust, are probably 4 to 5 times larger than man-made ones on a global scale, but regional variations in man-made pollution may change this ratio significantly in certain areas, particularly in the industrialised Northern Hemisphere. Although making up only 1 part in a billion of the mass of the atmosphere, they have the potential to significantly influence the amount of sunlight reaching the Earth's surface, and therefore climate.

Introduction

Removal of most aerosols is mainly achieved by rainfall (wet deposition) and by direct uptake at the surface (dry deposition). Explosive volcanic eruptions however, can inject large quantities of dust and gaseous material, such as sulphur dioxide, high into the atmosphere (the stratosphere). Here, sulphur dioxide is rapidly converted into sulphuric acid aerosols. Whereas pollution of the lower atmosphere is removed within days by the effects of rainfall and gravity, stratospheric pollution may remain there for several years, gradually spreading to cover much of the globe.

Like greenhouse gases, aerosols influence the climate. Atmospheric aerosols influence the transfer of energy in the atmosphere in two ways: directly through the scattering of sunlight; and indirectly through modifying the optical properties and lifetimes of clouds. The scattering of sunlight by aerosols is clearly demonstrated in the aftermath of a major volcanic eruption, when exceptionally colourful sunsets may be witnessed. The volcanic pollution results in a substantial reduction in the direct solar beam, largely through scattering by the highly reflective sulphuric acid aerosols. Overall, there is a net reduction of 5 to 10 per cent in energy received at the Earth's surface. An individual eruption may cause a global cooling of up to 0.3°C, with the effects lasting 1 to 2 years.

Estimation of the impact aerosols have on longer-term global climate change however, is more complex and hence more uncertain than that due to the well-mixed greenhouse gases. This is largely because the geographical distribution of aerosols is highly variable and strongly related to their sources. The best estimates of global cooling attributable to man-made aerosols are based on computer models. These show that the global cooling effect of man- made aerosols could offset the warming effect of increased greenhouse gas concentrations by as much as 30 per cent. The variable distribution of aerosols however, makes calculation of a global average difficult. Nevertheless, it is likely that aerosols may slow the rate of projected global warming during the 21st century.

AGRICULTURE

Climate is the most significant factor in determining plant growth and productivity. Without intervention to reduce emissions of greenhouse gases, global average surface temperature is projected to increase by about 0.2°C per decade during the 21st century. This rapid change in climate will have major implications for agriculture around the world. Crop growth is often limited by temperature. Temperatures during the 21st century are expected to increase more in the higher latitudes where shifts in vegetation will be greater. In Britain an increase in temperature of 1.5°C by 2050 is the equivalent of a decrease in altitude of approximately 200m.

This is the same as a shift southward in latitude of 200-300 km. Such an increase in temperature would allow widespread maize cultivation across southern England to take place. In other regions however, a rise in temperature may not be so beneficial. Small increases in temperature would extend the range of temperature-limited pests. The European Corn Borer for example, a major pest of grain maize, may shift between 165 and 500 km northwards with a rise of 1°C.

Moisture and water availability will be affected by a temperature increase, regardless of any change in rainfall. Higher temperatures increase the evaporation rate, thus reducing the level of moisture available for plant growth, although other climatic elements are involved. A warming of 1°C, with no change in precipitation, may decrease yields of wheat and maize in the core cropping regions such as the US by about 5 per cent. A very large decrease in moisture availability in the dryer regions of the world would be of great concern to the subsistence farmers that farm these lands. Reduced moisture availability would only exacerbate the existing problems of infertile soils, soil erosion and poor crop yields. In the extreme case, a reduction in moisture could lead to desertification.

Sea levels have been projected to rise by anywhere up to a metre by 2100, although considerable uncertainty is attached to this. The greatest threat to low-lying agricultural regions from sea level rise is that of inundation and flooding. Southeast Asia would have the greatest threat of inundation because of the deltaic nature of the land. Furthermore, the pollution of surface and groundwater with salty seawater is another potential problem facing farmers situated in low-lying regions. The costs of agricultural production would increase, resulting in higher food prices for the consumer. Although climate changes may have some detrimental impacts on agricultural production around the world, the increase in atmospheric carbon dioxide concentrations could be beneficial. Plants grow as a result of photosynthesis - the mechanism whereby the plant converts carbon dioxide from the atmosphere into food. With higher levels of carbon dioxide stimulating the rate of photosynthesis, the growth rate and productivity of plants could be expected to increase. This would be beneficial for global food stocks.

Most crops grown in cool, temperate regions respond positively to an increased concentration of carbon dioxide, including some of the current major food staples such as wheat, rice and soybean. Some studies have shown that growth rate in these crops may increase up to 50 per cent if carbon dioxide in the atmosphere is doubled. Crops grown in the tropical regions of the world, including sorghum, maize, sugar cane and millet, which together account for about one fifth of the world's food production, do not respond as well to increases in carbon dioxide. In order to maintain agricultural output to meet the demand for a growing world population, farmers will have to adjust and adapt as and when necessary to the possible changes imposed by changing climate. Higher temperatures would increase the demand for irrigation of agricultural land.

Unfortunately, in many arid and semi-arid regions of the world the demand for water already exceeds supply. Increased spread of pests and disease may also place additional demands on the need for fertilizers, pesticides and herbicides which are costly. The ability to adapt to the effects of climate change will vary greatly between countries and regions. Economic and technological constraints will limit the rate of adaptability, with poorer economies lagging behind. Consequently, without intervention the effects of climate change in the 21st century look set to further widen the gulf between developed and developing nations.

EVIDENCE FOR GLOBAL WARMING

Degradation of Earth's Atmosphere; Temperature Rise; Glacial Melting and Sealevel Rise; Ozone Holes; Vegetation Response "Global change". "Greenhouse effect". "Global warming". The media are full of statements, concerns, guesses, and speculation about these phenomena, as scientists and policy-makers around the world struggle to address recent scientific observations that indicate human activities impact our environment. And yet, each of these is a "natural" phenomenon, as are many others. Hurricanes, droughts, and monsoons all occur without any control by humans, to initiate, forestall or moderate them. Most readers of this Tutorial already know that global warming is litreally a "hot" topic. For those still with relative unfamiliarity, we recommend this Wikpedia review: effects of global warming.

This page considers the nature of and evidence for global warming. There seems almost no doubt that the Earth's climate has been steadily warming overall in the last hundred years: global warming is a fact - a reality. The nagging question - still unanswered to everyone's satisfaction - is: How much of the quantified increase of about 1° C in that time interval is just due to a natural trend probably related to glacial cycles (which themselves are temperature-dependent) and how much is additional warming accelerated by human activity as the industrial revolution reaches rapidly increasing levels. The present page provides an overview. Since it was written, the writer (NMS) has encountered new lines of evidence - pro and con.

Most people who follow the news accounts of global warming have come to realise that the primary culprit is human contributions of heat-affecting gases into the atmosphere as these are released by everyday processes such as burning coal to produce electricity and consuming gasoline to run automobiles. The most blamed gas is carbon dioxide (CO_2). The United States is often cited as the principal polluting nation, although now China and India - each with more than a billion people, many of whom are now driving cars - are becoming major players. In the U.S., cities and surrounding metropolitan areas are the chief sources of CO_2, as evident in this map made from ground and aerial sounding monitors but also containing some input from satellite measurements:

To fully understand global warming, one must first understand the operations of the Earth System, since natural sources of temperature-affecting gases play an important role. We can learn about our planet's interacting physical systems by observing the results of such natural phenomena, and use our knowledge to explore human-induced changes. Thus, one factor that may in itself induce and account for some of the regional and global temperature changes is volcanism. Remote sensing is effective at monitoring such events. Consider, for example, the eruption of a volcano, such as Mount

Pinatubo in the Philippine Islands in 1991, that happened without human intervention. This volcano had been dormant for more than 600 years.

When a volcano erupts it spews millions of tons of ash, debris, and gases into the atmosphere, not to mention the lava flows from some volcanoes. Because of the presence of instruments - on the ground, at the ocean's surface, and in space - meteorologists/environmentalists observed a cloud of sulfur dioxide (SO_2), emitted by Pinatubo, make its way westward, extending well past India within twelve days of the original eruption.

Monitoring has been done by the UARS (Upper Atmosphere Research Satellite). By three months, that cloud had completely encircled the Earth, as shown from space (below), and inside of a year SO_2 particles in the atmosphere were providing gloriously-coloured sunsets all over the globe and lowering global temperatures, as well. Clearly, an erupting volcano impacts more people and places than just within its immediate vicinity.

Volcanoes are a main source of SO_2 rises in atmospheric chemistry. The TOMS instrument has been making measurments of SO_2 in the air for more than 30 years now. Another fascinating example of a natural phenomenon we know as El Niño, because it occurs with some regularity (although not complete predictability) around Christmas time. El Niño refers to the baby Jesus, whose birthday we celebrate at the end of December. When an El Niño occurs, a pool of warm water from the western Pacific Ocean moves eastward

Introduction 17

to the western coast of South America. In the process, weather patterns around the world changes often to the detriment of human populations– as do South American fish populations. In non-El Niño years, fish are abundant in this region, because of the cold, nutrient-filled waters. When an El Niño occurs, that cold water flows deep into the Pacific, and fish populations decline dramatically, with concomitant effects on humans whose livelihood depends on those fish.

Other such phenomena abound. Some are readily observable by space sensors, particularly meteorological and oceanographic measurements. As examples, consider these near-global plots. The top from comes from the Seasat Radar Scatterometre and shows prevailing wind patterns over the oceans at supercontinental scales. The bottom one depicts the mean day, night, and day- nite temperatures of the Earth's land and sea surfaces, averaged for January of 1979, from Nimbus 6's High-resolution Infra-red Sensor (HIRS) 3.4 and 4.0 ìm channels, integrated with MSU microwave and infrared data.

Suffice to say that even without human contributions Earth is a dynamic system, one that changes routinely and often drastically. Clearly, global mean annual temperatures are rising, and we continue to monitor this condition with space observations to help settle the question: how much is just a natural trend (e.g., inevitable interglacial warming) and how much is due to man's activities?

Calculations show that the burning of fossil fuels (mainly coal, petroleum derivatives, and natural gas) add about 6 billion metric tons of carbon (as the element) to the air annually; each year also, deforestation permits an extra 1- 2 billion metric tons of carbon to reach the atmosphere. As an indication of how much worldwide temperatures have risen in the last few years, this next map of the globe shows the geographic distribution of temperature anomalies, with measurements from various sources as compiled by NASA GISS (Goddard Institute for Space Science), for the year 2006 - the fifth warmest on record.

The plot of temperature increase since 1880 is shown but is hard to read. So, beneath it is a recent version that clearly presents details: GISS is continually refining its models. Daily inputs improve the credibility of the predictions. Now that Aqua is operational, its AMSR instrument can provide global data covering very short time spans. This map is of worldwide surface temperatures on August 27, 2003:

Using temperature data, GISS has now published a map of sea surface heat content anomalies (measured to a water depth of 70 m) integrated over a ten year period from 1993 to 2003. From typical data they have derived a model distribution of temperature variations. Since the upper ocean waters serve as efficient reservoirs for heat storage, the patterns of distribution of excess, normal, and deficient heat (in Watts-year/m2) thus determined are proving helpful in weather and climate forecasting since the temperatures have strong effects on atmospheric heating.

The trend of temperature changes on annual and 5 year-running averages proves that in the second half of the 20th Century the global atmosphere has been slowly warming - cumulatively totaling about 0.5 ° C, as shown here:

Under the 2085 conditions, rainfall in this region will diminish and vegetation will become much like that around Phoenix, AZ today. Available water will become scarce, so that desalination of the Atlantic ocean would have to provide much that is needed.

That this warming trend is already happening elsewhere in North America is convincingly demonstrated by temperature data for the state of California. In the illustration, temperatures (in Fahrenheit) across the state and in Los Angeles have risen more than 2.5° F in places in the last four decades and more generally almost 4° F statewide in the last 120 years:

So, is there any sign this is happening everywhere? Much debate has started in the middle of 2006 because of very intense heat waves during July-August in both the U.S. and Europe (the latter seldom experiences the extremes that occurred then) as shown in these plots obtain from metsats: These excess temperatures point up the problem that both advocates and disbelievers of global warming face in interpreting the abnormalities. Thus: Does this period of high temperatures (St. Louis, MO had almost a week of temperatures at or above 100° F) reflect the onset of the predicted hotter summers or is this just another of the drastic warm spells that have happened infrequently in the past? The "jury is still out", although the unusual European situation seems to favour the warming hypothesis.

accumulated other relevant environmental recordings. For example, levels of several trace gases in our atmosphere have been rising and continue to rise. This change in atmospheric gas composition may be the most serious threat to the global well-being of Earth's environment.

Information about this has been nicely summarized in an October, 1997 pamphlet entitled *Climate Change: State of Knowledge*, prepared and distributed by the President's Office of Science and Technology (OST), from which information and selected illustrations on this page have been extracted. One of these gases, carbon dioxide (CO_2) has been increasing at an accelerating rate since the middle of the last century, after many, many years of essentially stable levels. Why is this? A simple explanation is that the Industrial

Introduction

Revolution began at about the time these increases started. With that social upheaval came the use of biomass and coal for fuel to support these new industries. Burning such material generates CO2.

Other trace gases have been rising, as well. Carbon Monoxide (CO) is particularly deleterious. Methane (CH4), from rice paddy production and enteric fermentation, is increasing, as are chlorofluorocarbons (CFCs) that have been used for many years as a refrigerant and to produce foam. Methane is a much more potent greenhouse gas that CO2.This pie chart shows the relative contributions of the different gases involved in atmospheric contamination:

As you will learn shortly, CO2 is generally considered to be the prime culprit in any supposed global warming. For the next few paragraphs, we will divert from this theme to review some basic facts that provide a background for discussing CO2's role in moderating the atmosphere:

As we saw subsection that presented a "short course" in meteorology, the Earth's atmosphere obtains almost all its thermal energy from the Sun. This diagram is based on the general blackbody radiation curve.It shows the peak wavelength for radiation input from the Sun and the peak wavelength of the Earth itself as a thermal body of average temperature of 288 Kelvin. (The two curves have the same height in the plots; in fact the height for Earth is greatly subdued, indicating its energy output is much lower.) The amount of carbon dioxide added or subtracted through natural processes.

This results in two kinds of plots

1. One that shows variations (ups and downs) over long periods; and
2. One that shows a steady increase in recent years.

This is pertinent information but a third plot would be equally informative: one that simply shows how temperature increases with concentration (the amount of CO2 in a fixed volume). This is a straightforward type of measurement in Physics.

The irrefutable message from this plot: experimentally, it is clearly demonstrated that the addition of increasing amounts of CO2 to an otherwise constant atmosphere causes a systematic rise in gas temperature. The ranges shown in the above plot are pertinent to levels that might be attained in the future if carbon dioxide emissions are not curbed.

Now, back to a survey of information relevant to global warming. Later we will familiarize you with dedicated satellites that gather data on atmospheric gases and other contributors to global warming. This bears on a main thesis in this Tutorial, that satellites are proving to be a powerful, almost unique, tool for monitoring the whole Earth at the global scale.

These new data sets will be invaluable. Scientists need to establish a long-term data base for global warming, which includes temperatures from previous centuries. One way to estimate past temperatures is from CO_2 measurements of air trapped in glacial ice that has been accumulating for thousands of years.

Mankind can probably live with these changes as we adjust and find ways to accommodate them. But the Gore book describes some climate-influenced events that will be devastating and considers certain possible effects that may be catastrophic. The many varied observations described on this page seem to favour the global warming hypothesis. But we must reiterate one statement made above: This warming may just be (largely) a natural effect of an interglacial temperature process (but nevertheless there clearly seems to be a humans' adverse contributions). If so, not to worry excessively (but some protective response is in order); if not, then the Gore conclusion - that very serious damage to global environments is about to occur - should be a warning that seems to cry out for a quick and drastic response by the planet's polluting nations.

In early 2007, a panel of more than a thousand scientists released a pair of reports summarizing what Science can state with high reliability about global warming.

To paraphrase some of their conclusions:

- Over the past few million years, natural climate changes, driven mainly in response to factors that cause widespread glaciation and alternate interglacial periods, are marked by intervals of slowly rising temperatures (interglacials) and then falling temperatures (continental glaciation). Rates of change are usually less than 1 degree per thousand years, and the total range is probably less than 20° F.
- During interglacials, some animal or plant species adapted to colder weather have died off (the same may happen during warm weather; new species can develop during either cold or warm climes).
- During interglacials, sealevel tends to rise and vegetation distribution varies; many biomes migrate poleward; productivity of edible plants may increase or decrease.
- It is likely that mankind has blossomed since the end of the last glaciation and has become both numerous and dispersed in the last ten thousand years. Now humans are in such abundance as to begin to affect (impact) natural conditions that control the weather, moderate the environment, and otherwise maintain healthy living conditions.

Introduction

- In all probability, some of the temperature rise whose rate is increasing is just a consequence of the natural interglacial warming. But a proportion of that rise (amount still guesswork) is almost certainly the influence of detrimental atmospheric gases that are being added from fossil fuel burning and other sources. The rise may be abnormal - it is the sum of both natural and manmade contributions.
- Since the human factor was not present in earlier interglacial warming trends, its current role in causing global warming cannot be evaluated in terms of outcomes. Many scientists believe that the rate and extent of temperature rise can be excessive enough to cause global equilibrium (adjustments of life to the changes) to be upset, perhaps to a stage that will lead to catastrophe. Work must be done to predict what might happen. mankind plays it safe and reduces or eliminates the known causes of atmospheric warming.

DETECTION OF ANTHROPOGENIC GLOBAL WARMING

The anthropogenic increase in atmospheric concentrations of greenhouse gases, and the associated increase in greenhouse radiative forcing. examined the observed climate changes, principally in surface temperature, that have taken place during the last 100 to 140 years. In this, the evidence for a causal link between the anthropogenic increase in greenhouse radiative forcing and the observed global warming will be reviewed.

The word "detection" by climate scientists has been used to refer to the identification of the significant change in climate during the twentieth century and its association with the anthropogenically enhanced greenhouse effect. Until 1995 most reviews had concluded that the enhanced greenhouse effect has not yet been detected unequivocally in the observational record. However, they have also noted that the global-mean temperature change over the past 100 years is consistent with the greenhouse hypothesis. Since then, the most recent scientific report from the Intergovernmental Panel on Climate Change has proposed that the balance of [modelling] evidence suggests a discernible human influence on the global climate does exist.

Traditionally, the difficulty in greenhouse detection has arisen because there are numerous other causes of climatic variability, and some of these [forcing mechanisms] may be operating on time scales (10^1 to 10^2 years) comparable to that of the anthropogenic greenhouse forcing. In addition, Wigley and Raper have shown that the inherent variability (random fluctuations) of the climate system can produce warming or cooling trends of up to 0.3°C per century.

The detection problem can be conveniently described in terms of "signal" and "noise". The signal here is the time-dependent climatic response to greenhouse forcing, whilst the noise is any non-greenhouse climate variation, either periodic, quasi-periodic or random.

ATTRIBUTION AND THE FINGERPRINT METHOD

In the last part, it was recognised that although a good case for linking the enhanced greenhouse effect to global warming could be proposed, a high degree of confidence could not be attached to such a cause-and-effect relationship from studies of a single variable (surface temperature). Linking cause and effect is referred to as attribution. Confidence in the attribution is increased as model predictions of changes in various components of the climate system are borne out by the observed data in more and more detail. This method is known as the fingerprint approach; namely, identification of an observed signal that has a structure unique to the predicted enhanced greenhouse effect. It was noted that the use of global-mean surface temperature as a fingerprint variable does not permit attribution.

The fingerprint method is essentially a form of model validation, where the perturbation experiment that is being used to test the models is the currently uncontrolled emissions of greenhouse gases into the atmosphere. In addition to global-mean surface temperature, there are a number of other fingerprint variables that can be used. In choosing them, three issues must be considered:
- The signal-to-noise ratio - should be maximised;
- Uncertainties in both the predicted signal and the noise - these should be minimised, and;
- The availability of suitable observational data.

LATITUDINAL SURFACE TEMPERATURES

Most simulations suggest that the warming north of 50°N in the winter half of the year should be enhanced due to ice-albedo feedback mechanisms. Over the last 100 years, high northern latitudes have warmed slightly more than the global mean during winter, but since the 1920s there is little noticeable trend. More importantly, however, high-latitude warming does not necessarily represent a unique fingerprint of greenhouse warming. Ice-albedo feedbacks could be equally invoked for other causes of climate change.

TROPOSPHERIC WARMING AND STRATOSPHERIC COOLING

All equilibrium model simulations show a warming in the mid-troposphere and cooling in the stratosphere. It has been suggested that this useful detection fingerprint. In addition, high signal-to-noise ratios have been obtained for free tropospheric temperatures. However, stratospheric cooling may not solely be attributed to greenhouse forcing, and can arise due to volcanic pollution injections and ozone depletion. Model simulations are also inconsistent with recent observations, in particular at the level at which warming reverses to cooling. Perhaps most significantly of all, though, is the lack of reliable instrumental data. The instrumental record for upper atmosphere temperatures extends back only to the late 1950s with the use of radiosonde, whilst satellite-based data is even younger.

Global-Mean Precipitation Increase

GCM models suggest an increase in global-mean precipitation, as one might expect from the associated increase in atmospheric temperature. However, because the spatial variability of rainfall is much greater than for surface temperature, regional and local details of changes are highly uncertain. Instrumental data from which long-term changes in precipitation can be determined are only available over land areas data suffers from the problems of incomplete coverage and inhomogeneity.

With the likelihood that the precipitation signal-to-noise ratio is low, any meaningful comparison between observation and model, and therefore attribution, is precluded. In addition, since global precipitation is partially dependent upon global temperature, rises in global precipitation may be expected as a consequence of other causes of climate change.

Sea Level Rise

Increasing greenhouse gases are expected to cause a rise in the global-mean sea level, due partly to oceanic thermal expansion and partly to the melting of land-based ice masses. However, as for global precipitation, such a variable is not wholly independent of global temperature. Thus, whilst both thermal expansion and glacial melting are consistent with global warming, neither provides any independent information about the cause of the warming.

MULTIVARIATE FINGERPRINTS

Although most univariate detection methods of global climate change due to an enhanced greenhouse effect have their limitations, it is likely that multivariate fingerprint methods, which involve the simultaneous use of several time series, will facilitate attribution. In its most general form one might consider the time evolution of a 3-D spatial field, comparing model results with observations. This may be achieved by comparing changes in mean values and variances, or correlating spatial patterns between simulation and observation

WHEN WILL ATTRIBUTION OCCUR

The fact that the cause (the enhanced greenhouse effect) and effect (global warming) have not unambiguously been linked leads to the question: when is this [attribution] likely to occur? Detection is not a simple yes/no issue. Rather, it involves the gradual accumulation of observational evidence in support of model predictions.

The scientist's task is to reduce the uncertainties associated with the understanding of the climate system. Thus, while the ultimate objective of climate research is detection, it is really the climate sensitivity one is interested in. A better understanding of this key parametre will not only allow one to argue with increased confidence that anthropogenic global warming is (or is not) occurring, but it will support prediction of future climate change, both anthropogenic and natural.

FUTURE CLIMATE CHANGE

The issue of climate change detection introduces the wider concern of future climate change. Although predictions of climate change thousand, millions and even hundreds of millions of years from now could be made based on our knowledge of longer term climate changes, I will be concerned only with the current decadal to century time scale. Clearly, prediction of climate change over the next 100 to 150 years is based solely on model simulations.

effects of continued anthropogenic pollution of the atmosphere by greenhouse gases, and to a lesser extent, atmospheric aerosols. The main concern, at present, is to determine how much the Earth will warm in the near future.

GCM CLIMATE SIMULATIONS

During the last decade or so, a number of complex GCMs have attempted to simulate future anthropogenic climate change. Earlier models studied equilibrium climate change associated between two, presumably stable, states of climate. The results of these are discussed in the first IPCC report. Associated

with a doubling of pre-industrial atmospheric CO2, the following conclusions have been made:
- A global average warming at or near the Earth's surface of between 1.5 and 4.5°C, with a "best guess" of 2.5°C, will occur;
- The stratosphere will experience a significant cooling;
- Surface warming will be greater at high latitudes in winter, but less during the summer;
- Global precipitation will increase by 3 to 15 per cent;
- Year-round increases in precipitation in high-latitude regions are expected, whilst some tropical areas may experience small decreases. More recent time-dependent GCMs which couple the atmospheric and oceanic components of the climate system together, provide more reliable estimates of greenhouse-gas-induced climate change. For steadily increasing greenhouse forcing, the global rise in temperature is typically less than the equilibrium rise corresponding to an instantaneous forcing. Significant results indicate:
- A global average warming of 0.3°C per decade, assuming non-interventionist greenhouse gas emission scenarios;
- A natural variability of about 0.3°C in global surface air temperature on decadal time scales;
- Regional patterns of temperature and precipitation change similar to equilibrium experiments, although warming is reduced in high latitude oceans where deep water is formed.

The ability to model the time-dependent nature of the climate system more adequately has allowed scientists to investigate the damping effects of the oceans on climate change.

Because the response time of the oceans, in particular the deep ocean, is much longer than for the free atmosphere, they have a regulating or delaying effect on the warming associated with enhanced greenhouse forcing. In addition, the transient GCMs have allowed increased attention to focus on the critical role of feedback processes in determining the climate's response to forcing perturbations.

GREENHOUSE FEEDBACKS

GCMs have estimated the CO2 doubling temperature change in the absence of feedback processes to be approximately 1.2°C. The existence of feedback loops within the climate system results in a climate sensitivity of 1.5°C to 4.5°C, and makes an otherwise linear association between radiative forcing and global temperature distinctly non-linear. Three of the most important direct climatic feedbacks to greenhouse forcing include the water vapour feedback, the cloud feedback and the ice-albedo feedback.

WATER VAPOUR FEEDBACK

The importance of water vapour feedback in climate change has long been recognised. The concentration of water vapour in the atmosphere increases rapidly with rising temperature (about 6 per cent/°C); this is the basis for the strong positive water vapour feedback seen in current climate models (whereby increases in temperature produce increases in atmospheric water vapour which in turn enhance the greenhouse forcing leading to further warming). All current GCMs simulate a strong positive water vapour feedback and are in agreement with observational data.

CLOUD FEEDBACK

Cloud feedback is the term used to encompass effects of changes in cloud and their associated radiative properties on a change of climate, and has been identified as a major source of uncertainty in climate models. This feedback mechanism incorporates both changes in cloud distribution (both horizontal and vertical) and changes in cloud radiative properties (cloud optical depth and cloud droplet distribution); these are not mutually independent. Although clouds contribute to the greenhouse warming of the climate system by absorbing more outgoing infrared radiation (positive feedback), they also produce a cooling through the reflection and reduction in absorption of solar radiation (negative feedback).

It is generally assumed that low clouds become more reflective as temperatures increase, thereby introducing a negative feedback, whilst the feedback from high clouds depends upon their height and coverage and could be of either sign.

ICE-ALBEDO FEEDBACK

The conventional explanation of the amplification of global warming by snow and ice feedback is that a warmer Earth will have less snow cover, resulting in a lower global albedo and consequent absorption of more radiation, which in turn causes a further warming of the climate. Most GCMs 1990, 1991) have simulated this positive surface albedo feedback, but significant uncertainties exist over the size of the effect, particularly with sea-ice.

GREENHOUSE GAS FEEDBACKS

In addition to these climatic feedbacks, there are other non-climatic feedbacks which may enhance or diminish the increase in atmospheric

Introduction

greenhouse gas accumulation. Potentially, there are numerous feedback processes which may act on the carbon cycle, thus affecting the transfer of carbon dioxide between the various components. Carbon storage in the oceans will be affected by both changes in water temperature and ocean circulation resulting from global warming. Warmer water stores less CO_2, and future atmospheric CO_2 increase may be amplified by something like 5 to 10 per cent by this effect. A slow down in vertical overturning within the oceans will reduce the uptake of atmospheric CO_2. In addition, reduced nutrient recycling would limit primary biological productivity in the surface oceans, thereby increasing further the surface water partial pressure of CO_2, and consequently atmospheric CO_2.

Carbon cycle feedbacks involving terrestrial carbon storage may include carbon dioxide fertilisation of plant photosynthesis (a negative feedback); eutrophication involving nitrogen and phosphate nutrient fertilisation (a negative feedback); temperature feedbacks on the length of growing season and photosynthesis/respiration rates (both positive and negative feedbacks) and changes in the geographical distribution of vegetation. The methane gas cycle may also experience feedback effects. Recent attention has been focused on northern wetlands and permafrost, where increases in temperature and soil moisture would result in significant increases in methane release into the atmosphere.

Despite the increased ability of transient GCMs to model climate feedback processes, there is no compelling evidence to warrant changing the equilibrium sensitivity to doubled CO_2 from the range of 1.5 to 4.5°C as estimated by IPCC, 1990a.

GREENHOUSE MODELLING VERSUS OBSERVATION

Global-mean temperature has increased by around 0.3 to 0.6°C over the past 100 years. At the same time, greenhouse gas concentrations, and atmospheric aerosol loadings have increased substantially. To assess whether the two are associated requires the use of model simulations of the likely climatic effects of the changing atmospheric composition, and the comparison of the results with observations. Three important detection experiments will be discussed here. Wigley and Barnett (1990) used an energy balance climate model (incorporating upwelling and diffusion within the oceans to account for their radiative damping effect). The model was forced from 1765 to 1990 using only the changes in greenhouse gas concentrations, and the response could be varied by changing the value of the climate sensitivity † . In this way, climatic feedbacks, not explicitly modelled, can be incorporated into the model.

The model results were qualitatively consistent with the observations on the century time scale. On shorter time scales, the model failed to reproduce the inter-decadal variability of the instrumental record.

Indeed, this caveat has often been used as an argument against the greenhouse hypothesis altogether. However, Wigley and Barnett point out that such variability represents the background noise against which the greenhouse signal has to be detected. Significantly, the observational record seemed to lie at the low climate sensitivity end of the output range of that predicted by GCMs (1.5 to 4.5°C). However, the situation becomes more complex if other forcing mechanisms, in addition to the enhanced greenhouse effect, are invoked. If the net century time scale effect of non-greenhouse factors (e.g. solar variability, volcanism) involved a warming, the climate sensitivity would be less than 1°C. If their combined effect were a cooling, the sensitivity could be larger than 4°C.

One possible explanation for the decadal time scale discrepancies between the model and observed data is that some other forcing mechanism has been operating which has either offset or reinforced the general warming trend at different times. Using another energy balance model, Kelly and Wigley considered solar variability as a possible candidate. The model was run with a series of sensitivities spanning the accepted range of uncertainty in order to identify the best fit between modelled and observed temperature. Two sets of forcing histories (determined by a 1-D radiative-convective model) were considered, one involving only the effect of enhanced greenhouse gas concentrations (the IPCC 1990 forcing record), the other including also the negative radiative effect of aerosol loading and stratospheric ozone depletion.

Table summarises the results of Kelly and Wigley. The model explicitly calculated the best fit CO_2 doubling temperature (climate sensitivity) and the amount of explained variance in the observational record by each forcing history. As well as IPCC 1990 and 1992 forcing histories, different solar variables (sunspot number, length on sunspot cycle, solar diametre and rate of change of solar diametre) were considered, and combined with the greenhouse forcing.

Table. Summary results of the Wigley/Kelly model

Forcing		Doubling	Explained variance (%)		
Green-house	Solar	Temp.(°C)	Green-house	Solar	Total
IPCC 90/92	None	1.8/3.8	46.1/49.2		46.1/49.2
IPCC 90/92	Number	1.2/2.9	39.7/46.1	10.7/6.4	50.4/52.5
IPCC 90/92	Length	0.9/1.9	30.4/33.8	22.6/19.8	53.0/53.6
IPCC 90/92	Diametre	1.1/2.6	37.1/43.6	16.4/10.4	53.5/54.0
IPCC 90/92	Gradient	1.8/3.6	46.1/49.0	8.9/8.0	55.0/57.0

Table demonstrates that a considerable difference in the best-fit climate sensitivities exists between the two IPCC forcing histories. If greenhouse forcing alone is considered the CO_2 doubling temperature is at the low end of the range predicted by GCMs, and is in agreement with Wigley and Barnett. If the negative forcing of aerosol loading and ozone depletion is also included, the climate sensitivity is much greater. This result is intuitively correct, since the radiative effects of aerosols and ozone loss would offset the warming due to increased greenhouse gases.

When solar variability is included into the model, the explained variance of the observational record is greater than for greenhouse forcing alone. This is true for all of the solar variables considered here. The forcing combination that explains the most variance in the observational record (57 per cent) includes the effects of greenhouse gases, aerosols and ozone depletion, and the rate of change of solar diametre. The latter, it seems, is accounting for much of the inter-decadal variability in the instrumental record.

There now seems to exist a plausible explanation for why the warming has not been regular over the past 100 years, but Kelly and Wigley point out a number of caveats. Most importantly, the explained variance varies little over a range of estimates, about the best-fit value, for the CO_2 doubling temperature.

Thus, it remains difficult to define a precise value for the climate sensitivity. Ultimately, knowledge of the climate sensitivity provides the key to projecting the future impact of greenhouse gas emissions. Whilst uncertainty remains, forecasting future climates will remain an imprecise science.

In addition, other possible mechanisms of climate forcing have not been considered. These could include the effects of volcanism, ocean circulation or even the inherent random variability of the climate system. However, as other potential mechanisms are included into the model, the uncertainty in the climate sensitivity rises. If sufficient data existed to define the natural level of climate variability prior to human interference, this estimate could be used to attach error margins to the analysis of the greenhouse signal without explicit consideration of the source of the background noise. Unfortunately, such data do not exit since the period of the instrumental record contains this greenhouse signal.

Another attempt to simulate the observational temperature record used a GCM which modelled the transient † response to anthropogenic radiative forcing. Being more computationally complex than energy-balance models, the Hadley Centre model could simulate both global-mean temperature changes and inter-regional differences. The model was run, incorporating the forcing histories of both greenhouse gases and aerosols.

For the first time, a GCM was able to replicate in broad terms the slow rise in global temperature since the middle of the last century. If greenhouse gases alone were influencing climate, one would expect global temperatures to have risen by some 0.6 to 1.3°C over the last 100 years. By taking into account the anthropogenic sulphate aerosols, the Hadley Centre model simulated a rise in temperature close to the observed 0.5°C.

What makes the experiment so interesting was that the GCM contained simulations of the atmosphere, oceans, ice and vegetation, and can therefore be considered to be a much better representation of the climate system than earlier GCMs, containing only the atmospheric component. Such a model significantly increased the confidence in scientists' assertion that current global warming is due to an anthropogenically enhanced greenhouse effect, albeit muted by the radiative effects of atmospheric aerosol loading. Indeed, it was in response to the Hadley Centre model, and similar ones to it more recently, that the Intergovernmental Panel on Climate Change indicated that the balance of [modelling] evidence suggests a discernible human influence on the global climate does exist. Nevertheless, in view of the reservations highlighted by Kelly and Wigley concerning the uncertainty of the climate sensitivity, it remains arguable whether the cause (enhanced greenhouse forcing) and effect (global warming) have been linked unequivocally.

Ozone Depletion

INTRODUCTION

Ozone is both beneficial and harmful to us. Near the ground, ozone forming as a result of chemical reactions involving traffic pollution and sunlight may cause a number of respiratory problems, particularly for young children, However, high up in the atmosphere (19-30km) in a region known as the stratosphere, ozone filters out incoming radiation from the Sun in the "cell damaging" ultraviolet part of the spectrum. Without ozone in the stratosphere, life on earth would not have evolved. Thus with the development of the ozone layer came the formation of more advanced life forms. Concentrations of ozone in the stratosphere fluctuate naturally in response to variations in weather conditions and amounts of energy being released from the sun, and to major volcanic eruptions. Nevertheless, during the 1970s it was realised that man- made emissions of CFCs and other chemicals used in refrigeration, aerosols and cleansing agents may destroy significant amounts of ozone in the stratosphere, thereby letting through more of the harmful ultraviolet radiation.

Then, in 1985, a large "ozone hole" was discovered the continent of Antarctica during the springtime. This has reappeared every year. In response to this, and additional fears about more widespread global ozone depletion, 24 nations signed the Montreal Protocol on Substances that Deplete the Ozone Layer (1987). This legally binding international treaty called for participating developed nations to reduce the use of CFCs and other ozone depleting substances. In 1990 and again in 1992, subsequent amendments to the Protocol brought forward the phase out date for CFCs for developed countries to 1995. Protecting the ozone layer is essential. Ultraviolet radiation from the Sun can cause a variety of health problems in humans, including skin cancers, eye cataracts and a reduction in the ability to fight off disease. Furthermore, ultraviolet radiation can be damaging to microscopic life in the surface oceans which forms the basis of the world's food chain, certain varieties of crops including rice and soya, and polymers used in paints and clothing.

A loss of ozone in the stratosphere may even affect the global climate. International agreements have gone a long way to safeguarding this life-supporting shield. Nevertheless, for there to be real and longlasting success, everyone must become part of the solution. Individual efforts taken together can be powerful forces for environmental change. There are a number of things that we, as individuals, can do to both protect the ozone layer.

These include proper disposal of old refrigerators, the use of halon-free fire extinguishers and the recycling of foam and other non-disposable packaging. Finally, we should all be aware that whilst emissions of ozone depleters are now being controlled, the ozone layer is not likely to fully repair itself for several decades. Consequently, we should take precautions when exposing ourselves to the Sun. The Atmospheric Research and Information Centre (aric), through its Atmosphere, Climate and Environment Information Programme, has compiled a series of 20 topical fact sheets concerning the subject of ozone depletion. The series is divided into three parts - the science of ozone depletion, the impacts of ozone depletion, and managing ozone depletion.

Together, they describe what ozone, the ozone layer, ozone depletion and the ozone hole are, how ozone depletion occurs, mankind's influence, its impacts, and the international agreements put in place to control it. The fact sheet series is aimed at students involved in Key Stage 4 of the National Curriculum (GCSE) and higher. Although some of the concepts covered by the fact sheets may be challenging which sometimes contains words and phrases that may seem unfamiliar to the reader. aric hope that the reader will find this fact sheet series a useful information resource on ozone depletion.

WHAT CAUSES OZONE DEPLETION?

It was first suggested, by Drs. M. Molina and S. Rowland in 1974, that a man-made group of compounds known as the chloro-fluorocarbons (CFCs) were likely to be the main source of ozone depletion. However, this idea was not taken seriously until the discovery of the ozone hole over Antarctica in 1985. Chlorofluorocarbons are not "washed" back to Earth by rain or destroyed in reactions with other chemicals. They simply do not break down in the lower atmosphere and they can remain in the atmosphere from 20 to 120 years or more. As a consequence of their relative stability, CFCs are instead transported into the stratosphere where they are eventually broken down by ultraviolet radiation, releasing free chlorine.

Ozone Depletion 33

The chlorine becomes actively involved in the process of destruction of ozone. The net result is that two molecules of ozone are replaced by three of molecular oxygen, leaving the chlorine free to repeat the process:

HOW LONG HAS OZONE DEPLETION BEEN OCCURRING?

Based on data collected since the 1950s, scientists have determined that ozone levels were relatively stable until the late 1970s. Severe depletion over the Antarctic has been occurring since 1979 and a general downturn in global ozone levels has been observed since the early 1980s.

HOW MUCH OF THE OZONE LAYER HAS BEEN DEPLETED AROUND THE WORLD?

Global ozone levels have declined an average of about 3 per cent between 1979 and 1991. This rate of decline is about three times faster than that recorded in the 1970s. In addition to Antarctica, ozone depletion now affects almost all of North America, Europe, Russia, Australia, New Zealand, and a sizeable part of South America. Short term losses of ozone can be much greater than the long term average. In Canada, ozone depletion is usually greatest in the late winter and early spring. In 1993, for example, average ozone values over Canada were 14 per cent below normal from January to April.

ORIGIN OF OZONE

The photochemical mechanisms that give rise to the ozone layer were discovered by the British physicist Sidney Chapman in 1930. Ozone in the Earth's stratosphere is created by ultraviolet light striking oxygen molecules containing two oxygen atoms (O_2), splitting them into individual oxygen atoms (atomic oxygen); the atomic oxygen then combines with unbroken O_2 to create ozone, O_3. The ozone molecule is also unstable (although, in the stratosphere, long-lived) and when ultraviolet light hits ozone it splits into a molecule of O_2 and an atom of atomic oxygen, a continuing process called the ozone-oxygen cycle, thus creating an ozone layer in the stratosphere, the region from about 10 to 50 km (32,000 to 164,000 feet) above Earth's surface. About 90 per cent of the ozone in our atmosphere is contained in the stratosphere. Ozone concentrations are greatest between about 20 and 40 km, where they range from about 2 to 8 parts per million. If all of the ozone were compressed to the pressure of the air at sea level, it would be only a few millimeters thick.

ULTRAVIOLET LIGHT AND OZONE

Although the concentration of the ozone in the ozone layer is very small, it is vitally important to life because it absorbs biologically harmful ultraviolet (UV) radiation coming from the Sun. UV radiation is divided into three categories, based on its wavelength; these are referred to as UV-A (400–315 nm), UV-B (315–280 nm), and UV-C (280–100 nm). UV-C, which would be very harmful to all living things, is entirely screened out by ozone at around 35 km altitude. UV-B radiation can be harmful to the skin and is the main cause of sunburn; excessive exposure can also cause genetic damage, resulting in problems such as skin cancer. The ozone layer is very effective at screening out UV-B; for radiation with a wavelength of 290 nm, the intensity at the top of the atmosphere is 350 million times stronger than at the Earth's surface. Nevertheless, some UV-B reaches the surface. Most UV-A reaches the surface; this radiation is significantly less harmful, although it can potentially cause genetic damage.

DISTRIBUTION OF OZONE IN THE STRATOSPHERE

The thickness of the ozone layer—that is, the total amount of ozone in a column overhead—varies by a large factor worldwide, being in general smaller near the equator and larger towards the poles. It also varies with season, being in general thicker during the spring and thinner during the autumn in the northern hemisphere. The reasons for this latitude and seasonal dependence are complicated, involving atmospheric circulation patterns as well as solar intensity.

Since stratospheric ozone is produced by solar UV radiation, one might expect to find the highest ozone levels over the tropics and the lowest over polar regions. The same argument would lead one to expect the highest ozone levels in the summer and the lowest in the winter. The observed behaviour is very different: most of the ozone is found in the mid-to-high latitudes of the northern and southern hemispheres, and the highest levels are found in the spring, not summer, and the lowest in the autumn, not winter in the northern hemisphere. During winter, the ozone layer actually increases in depth. This puzzle is explained by the prevailing stratospheric wind patterns, known as the Brewer-Dobson circulation. While most of the ozone is indeed created over the tropics, the stratospheric circulation then transports it poleward and downward to the lower stratosphere of the high latitudes. However in the southern hemisphere, owing to the ozone hole phenomenon, the lowest amounts of column ozone found anywhere in the world are over the Antarctic in the southern spring period of September and October.

Ozone Depletion

The ozone layer is higher in altitude in the tropics, and lower in altitude in the extratropics, especially in the polar regions. This altitude variation of ozone results from the slow circulation that lifts the ozone-poor air out of the troposphere into the stratosphere. As this air slowly rises in the tropics, ozone is produced by the overhead sun which photolyzes oxygen molecules. As this slow circulation bends towards the mid-latitudes, it carries the ozone-rich air from the tropical middle stratosphere to the mid-and-high latitudes lower stratosphere. The high ozone concentrations at high latitudes are due to the accumulation of ozone at lower altitudes.

The Brewer-Dobson circulation moves very slowly. The time needed to lift an air parcel from the tropical tropopause near 16 km (50,000 ft) to 20 km is about 4–5 months (about 30 feet (9.1 m) per day). Even though ozone in the lower tropical stratosphere is produced at a very slow rate, the lifting circulation is so slow that ozone can build up to relatively high levels by the time it reaches 26 km.

Ozone amounts over the continental United States (25°N to 49°N) are highest in the northern spring (April and May). These ozone amounts fall over the course of the summer to their lowest amounts in October, and then rise again over the course of the winter. Again, wind transport of ozone is principally responsible for the seasonal evolution of these higher latitude ozone patterns. The total column amount of ozone generally increases as we move from the tropics to higher latitudes in both hemispheres. However, the overall column amounts are greater in the northern hemisphere high latitudes than in the southern hemisphere high latitudes. In addition, while the highest amounts of column ozone over the Arctic occur in the northern spring (March–April), the opposite is true over the Antarctic, where the lowest amounts of column ozone occur in the southern spring (September–October).

OZONE DEPLETION

The ozone layer can be depleted by free radical catalysts, including nitric oxide (NO), nitrous oxide (N2O), hydroxyl (OH), atomic chlorine (Cl), and atomic bromine (Br). While there are natural sources for all of these species, the concentrations of chlorine and bromine have increased markedly in recent years due to the release of large quantities of man-made organohalogen compounds, especially chlorof-luorocarbons (CFCs) and bromofluorocarbons. These highly stable compounds are capable of surviving the rise to the stratosphere, where Cl and Br radicals are liberated by the action of ultraviolet light. Each radical is then free to initiate and catalyze a chain reaction capable of breaking down over 100,000 ozone molecules. The breakdown of ozone in the stratosphere results in the ozone molecules being unable to absorb ultraviolet radiation. Consequently, unabsorbed and dangerous ultraviolet-B radiation is able to reach the Earth's surface. Ozone levels, over the northern hemisphere, have been dropping by 4 per cent per decade. Over approximately 5 per cent of the Earth's surface, around the north and south poles, much larger (but seasonal) declines have been seen; these are the ozone holes. In 2009, nitrous oxide (N2O) was the largest ozone-depleting substance emitted through human activities.

Regulation

In 1978, the United States, Canada and Norway enacted bans on CFC-containing aerosol sprays that are thought to damage the ozone layer. The European Community rejected an analogous proposal to do the same. In the U.S., chlorofluorocarbons continued to be used in other applications, such as refrigeration and industrial cleaning, until after the discovery of the Antarctic ozone hole in 1985. After negotiation of an international treaty (the Montreal Protocol), CFC production was sharply limited beginning in 1987 and phased out completely by 1996.

On August 2, 2003, scientists announced that the depletion of the ozone layer may be slowing down due to the international ban on CFCs. Three satellites and three ground stations confirmed that the upper atmosphere ozone depletion rate has slowed down significantly during the past decade. The study was organized by the American Geophysical Union. Some breakdown can be expected to continue due to CFCs used by nations which have not banned them, and due to gases which are already in the stratosphere. CFCs have very long atmospheric lifetimes, ranging from 50 to over 100 years, so the final recovery of the ozone layer is expected to require several lifetimes.

Compounds containing C–H bonds have been designed to replace the function of CFC's (such as HCFC), since these compounds are more reactive and less likely to survive long enough in the atmosphere to reach the stratosphere where they could affect the ozone layer. While being less damaging than CFC's, HCFC's also have a significant negative impact on the ozone layer. HCFC's are also being phased out.

WHY IS THE OZONE LAYER IMPORTANT?

Ozone's unique physical properties allow the ozone layer to act as our planet's sunscreen, providing an invisible filter to help protect all life forms from the Sun's damaging ultraviolet (UV) rays. Most incoming UV radiation is absorbed by ozone and prevented from reaching the Earth's surface.

Ozone Depletion 37

Without the protective effect of ozone, life on Earth would not have evolved in the way it has.

OXYGEN AND OZONE

Ozone is a form of oxygen. The oxygen we breathe is in the form of oxygen molecules (O2) - two atoms of oxygen bound together. Ozone, on the other hand, consists of three atoms of oxygen bound together (O3). Most of the atmosphere's ozone occurs in the upper atmosphere called the stratosphere.

Ozone is colourless and has a very harsh odour. Normal oxygen, which we breathe, has two oxygen atoms and is colourless and odourless. Ozone is much less common than normal oxygen. Out of each 10 million air molecules, about 2 million are normal oxygen, but only 3 are ozone.

WHERE IS OZONE FOUND?

Most ozone is produced naturally in the upper atmosphere or stratosphere. While ozone can be found through the entire atmosphere, the greatest concentration occurs at an altitude of about 25 km. This band of ozone-rich air is known as the "ozone layer". The stratosphere ranges from an altitude of 10km to 50km and it lies above the troposphere (or lower atmosphere).

Stratospheric ozone shields us from the harmful effects of the Sun's ultraviolet radiation. Ozone also occurs in very small amounts in the troposphere (200 parts per billion). It is produced at ground level through a reaction between sunlight and volatile organic compounds (VOCs) and nitrogen oxides (NOx), some of which are produced by human activities such as driving cars.

Ground-level ozone is a component of urban smog, and can damage certain crops and trees, and can be harmful to human health. As there is increased solar radiation during the summer months, there tends to be a larger number of urban ozone episodes at this time. Ground level ozone is blown easily by surface winds to regions well beyond its source. As a pollutant it should not be confused with the separate problem of stratospheric ozone depletion.

Even though both types of ozone are exactly the same molecule, their presence in different parts of the atmosphere has very different consequences. Stratospheric ozone blocks harmful solar radiation - all life on Earth has adapted to this filtered solar radiation. Groundlevel ozone, in contrast, is simply a pollutant. It will absorb some incoming solar radiation, but it cannot make up for stratospheric ozone loss. This fact sheet series is concerned with stratospheric ozone depletion.

OTHER EFFECTS OF OZONE DEPLETION

As well as the effects to human health, land plants and aquatic life, which may occur as a consequence of ozone depletion, there are other impacts which could result from prolonged destruction of ozone in the stratosphere. These include damage to polymers used in buildings, paints and packaging, and changes in biogeochemical cycles affecting ground-level pollution (smog), acid rain and even climate change.

DAMAGE TO POLYMERS

Ozone depletion will cause many materials to degrade faster. These materials include PVC (used in window and door frames, pipes and gutters, etc.), nylon and polyester. They are all composed of compounds known as polymers. Synthetic polymers, naturally occurring biopolymers, as well as some other materials of commercial interest are adversely affected by solar UV radiation. Today's materials are somewhat protected from UV-B by special additives. Therefore, any increase in solar UV-B levels as a result of ozone depletion will therefore accelerate their breakdown, limiting how long they are useful outdoors. Shorter wavelength (i.e. more energetic) UV-B radiation is mainly responsible for photo-damage ranging from discolouration to loss of mechanical integrity in polymers exposed to solar radiation. The use of higher levels of conventional light stabilisers in polymerbased materials are likely to be employed to mitigate the effects of increased UV levels in sunlight. However, it is not certain how resistant such light stabilisers are themselves to increased levels of UV-radiation. In addition, their use will add to the cost of plastic products in target applications. With plastics rapidly displacing conventional materials in numerous applications, this is an important consideration particularly in the developing world. It is not certain yet how other materials, including rubber, paints, wood, document and textiles will be affected by increased UV radiation resulting from ozone depletion.

EFFECTS ON BIOGEOCHEMICAL CYCLES

Increases in solar UV radiation could affect terrestrial and aquatic biogeochemical cycles, thereby altering both sources and sinks of greenhouse and chemically-important trace gases *e.g.*, carbon dioxide (CO_2), carbon monoxide (CO), carbonyl sulphide (COS) and possibly other gases, including ozone. These potential changes would contribute to biosphere-atmosphere feedbacks that attenuate or reinforce the atmospheric build-up of these gases.

Ozone Depletion

Likely effects include an increase in smog in urban centres, and acid rain in rural areas.

EFFECTS ON CLIMATE

Whilst increases of UV radiation as a result of ozone depletion may affect the production and removal of carbon dioxide, the main greenhouse gas, ozone depletion itself can influence the global climate. Ozone is also a greenhouse gas, and as well as filtering out the incoming short-wave solar radiation, can absorb much of the outgoing long-wave terrestrial radiation (infrared radiation). If stratospheric ozone is destroyed, ozone's greenhouse effect is reduced and this could lead to a global cooling, offsetting some of the warming that may be occurring as a result of man-made emissions of carbon dioxide, methane and nitrous oxide. Ironically, when the ozone layer starts to repair itself in the next century as a result of a control on CFCs, this cooling potential will be lost. More significantly, the replacement chemicals to CFCs, the HCFCs, which themselves do not harm the ozone layer, are very strong greenhouse gases, and are further contributing to the potential problem of global warming.

MONTREAL PROTOCOL

In 1985 an international agreement, the Vienna Convention, was signed after three years of negotiating under the auspices of the United Nations Environment Programme. The Vienna Convention established mechanisms for international co-operation in research, monitoring, and exchange of data on emissions, on concentrations of CFCs and halons, and on the status of stratospheric ozone. It also set a framework for international negotiations on actual reductions of emissions. That same year, 1985, marked another seminal development in the evolution of scientific and public policy recognition of the stratospheric ozone issue - the discovery of the Antarctic ozone hole. On the basis of the Vienna Convention (1985), the Montreal Protocol on Substances that Deplete the Ozone Layer was negotiated and signed by 24 countries and by the European Economic Community in September 1987. The Protocol called for the Parties to phase down the use of CFCs, halons and other man-made halocarbons.

THE PROTOCOL

The Montreal Protocol represented a landmark in the international environmentalist movement. For the first time whole countries were legally bound to reducing and eventually phasing out altogether the use of CFCs and other ozone depleting chemicals. Failure to comply was accompanied by stiff penalties. The original Protocol aimed to decrease the use of chemical compounds destructive to ozone in the upper atmosphere by 50 per cent by the year 1999. The agreement was supplemented by agreements made in London in 1990 and in Copenhagen in 1992, by which the same countries promised to stop using CFCs and most of the other chemical compounds destructive to ozone by the end of 1995. The Protocol has been subsequently amended twice more, at Montreal in 1997 and at Beijing in 1999. In most cases it has been fairly easy to develop and introduce compounds and methods to replace CFC compounds.

CFC use in aerosols and foam plastic packaging has already been abandoned in most countries. On the other hand, compounds capable of replacing CFC compounds in cooling devices and insulating materials are still under development. In order to deal with the special difficulties experienced by developing countries it was agreed that they would be given

10 years grace, so long as their use of CFCs did not grow significantly. China and India, for example, are strongly increasing the use of air conditioning and cooling devices. Using CFC compounds in these devices would be cheaper than using replacement compounds harmless to ozone. An international fund has therefore been set up to help these countries to introduce new and environmentally more friendly technologies and chemicals. The depletion of the ozone layer is a world-wide problem which does not respect the frontiers between different countries. It can only be affected through determined international co-operation.

CAN OZONE DEPLETION BE RESERVED

Can the hole in the ozone layer be repaired? Yes. If concentrations of ozone-destroying chemicals are reduced, the natural balance between ozone creation and destruction can be restored. However, this might require the complete elimination of CFCs, halons, carbon tetrachloride, methyl chloroform, and methyl bromide. In late 1991, scientists estimated that even with the current global schedule to eliminate ozone-destroying substances, the ozone layer would not return to 'normal' (pre-1980 chlorine levels) until the middle of the 21st century. Nevertheless, the 1998 World Meteorological Organiszation Scientific Assesment of Ozone Depletion observed that abundance of ozone-depleting compounds in the lower atmosphere is now slowly declining from a peak in 1994.

PHASING OUT CFCS

The initial concern about the ozone layer in the 1970s led to a ban on the use of CFCs as aerosol propellants in several countries, including the

Ozone Depletion

U.S. However, production of CFCs and other ozone-depleting substances grew rapidly afterwards as new uses were discovered. Through the 1980s, other uses expanded and the world's nations became increasingly concerned that these chemicals would further harm the ozone layer. In 1985, the Vienna Convention was adopted to formalise international co-operation on this issue. Additional efforts resulted in the signing of the Montreal Protocol in 1987. After the original Protocol was signed, new measurements showed worse damage to the ozone layer than was originally expected.

In 1992, reacting to the latest scientific assessment of the ozone layer, the Parties decided to completely end production of halons by the beginning of 1994 and of CFCs by the beginning of 1996 in developed countries. Between 1986 and 1991, world-wide consumption of CFCs-11, CFC- 12 and CFC-113 decreased by 40 per cent, ahead of the schedule outlined in the Montreal Protocol and faster even than called for in the more ambitious 1990 London Amendment to the Montreal Protocol. Manufacturers, who were earlier convinced that CFCs were unique and irreplaceable, were finding themselves moving quickly to alternative processes and chemicals. Hydrocarbons have replaced CFCs as aerosol propellants and as blowing agents for foams, and as cleansing solvents in electronics manufacturing. HCFCs, which have much smaller ozone depleting potentials have replaced CFCs for refrigeration and air conditioning.

ARE INTERNATIONAL AGREEMENTS ENOUGH

Without the Montreal Protocol, continued use of CFCs and other compounds would have tripled the stratospheric abundance of chlorine and bromine by 2050. Because of measures taken under the Protocol, emissions of ozone-depleting substances are already falling. Under current agreements, the stratospheric concentrations of chlorine and bromine are expected to reach their maximum within a few years and then slowly decline, although concentrations of chlorine are already falling in the troposphere. With evidence that international agreements to phase out the use of ozone depleting chemicals appear to be working, both NASA (the North American Space Agency) and NOAA (the National Oceanographic and Atmospheric Administration) in the United States of America have expressed confidence that, all other things being equal, the stratospheric ozone layer should return to normal by the middle of the next century.

The recovery of the ozone layer will be gradual because of the long times required for CFCs to be removed from the atmosphere; some take as long as several hundred years. Nevertheless, the likelihood remains that deep ozone holes will continue to form annually in the polar regions, well into the next century. This situation will persist until stratospheric chlorine levels decrease. The Intergovernmental Panel on Climate Change (IPCC), even if the control measures of the 1990 London and 1992 Copenhagen Amendments were to be implemented by all nations, the abundance of stratospheric chlorine and bromine will increase over the next several years. The Antarctic ozone hole, caused by industrial halocarbons, will therefore recur each spring. In addition, since these gases are also responsible for the observed reduction in middle and high latitude stratospheric ozone, the depletion at these latitudes is predicted to continue unabated for at least 5 to 10 years.

IMPACT OF ODSS ON STRATOSPHERIC OZONE

The abundance of ozone in the stratosphere at a particular location is governed by three processes: photochemical production, destruction by catalytic cycles, and transport processes. Photochemical production in the stratosphere occurs mostly through the photolysis of O_2, with loss via catalytic cycles involving hydrogen, nitrogen, and halogen radicals. The relative importance of the various loss cycles in the stratosphere varies substantially with altitude. Above about 45 km, loss through HO_x dominates, while below this altitude NO_x-catalyzed ozone loss is most important. The importance of the ClO_x-catalyzed ozone loss cycle, which varies with chlorine loading, peaks at about 40 km. Below about 25 km HO_x-driven ozone loss cycles dominate again.

Upper Stratosphere

In the upper stratosphere the ozone budget is largely understood, although uncertainties remain regarding the rate constants of key radical reactions. In particular, strong evidence has been accumulated that the observed ozone depletion in the upper stratosphere is caused by increased levels of stratospheric chlorine as originally proposed by Molina and Rowland and Crutzen.

Because of the direct correspondence between the stratospheric chlorine abundance and ozone depletion in the upper stratosphere, it has been suggested that the response of stratospheric ozone to the declining stratospheric chlorine levels might be first detectable in the upper stratosphere.

Polar Regions

In recent decades stratospheric ozone losses have been most pronounced in polar regions during winter and spring.

Ozone Depletion

These losses are determined largely by three chemical factors:

1. The conversion of chlorine reservoirs into active, ozone-destroying forms through heterogeneous reactions on the surfaces of polar stratospheric cloud particles;
2. The availability of sunlight that drives the catalytic photochemical cycles that destroy ozone;
3. The timing of the deactivation of chlorine. emperature controls the formation and destruction of polar stratospheric clouds and thus the timing of activation and deactivation of chlorine.

Furthermore, temperature controls the efficiency of the catalytic cycles that destroy ozone in the presence of sunlight. However, polar ozone loss would not occur without a prominent dynamical feature of the stratosphere in winter and spring: the polar vortex. In both hemispheres, the polar vortex separates polar air from mid-latitude air to a large extent, and within the vortex the low-temperature conditions that develop are the key factor for polar ozone loss. These two factors are dynamically related: a strong vortex is generally also a colder vortex. The two crucial questions for future polar ozone are whether, in an increased greenhouse-gas climate, the region of low stratospheric temperatures will increase in area and whether it will persist for longer in any given year.

When anthropogenically emitted ODSs are eventually removed from the stratosphere, the stratospheric halogen burden will be much lower than it is today and will be controlled by the naturally occurring source gases methyl chloride and methyl bromide. Under such conditions, dramatic losses of polar ozone in winter and spring as we see today are not expected to occur. However, the rate of removal of anthropogenic halogens from the atmosphere depends on the atmospheric lifetimes of CFCs and is considerably slower than the rate at which halogens have been increasing in the decades prior to approximately the year 2000, when the stratospheric halogen loading peaked. A complete removal of anthropogenic halogens from the atmosphere will take more than a century. During this period of enhanced levels of halogens caused by past anthropogenic emissions, the polar stratosphere will remain vulnerable to climate perturbations, such as increasing water vapour or a cooling of the stratosphere, that lead to enhanced ozone destruction.

Lower-Stratospheric Mid-Latitudes

It is well established that ozone in the lower stratosphere at midlatitudes has been decreasing for a few decades; both measurements of ozone locally in the lower stratosphere and measurements of column ozone show a clear decline. However, because mid-latitude ozone loss is much less severe than polar ozone loss, it cannot be identified in measurements in any one year but is rather detected as a downward trend in statistical analyses of longer time- series. It is clear that chemical loss driven by halogens is very important, but other possible effects that may contribute to the observed mid-latitude trends have also been identified. A definitive quantitative attribution of the trends to particular mechanisms has not yet been achieved. WMO reviewed this issue most recently; as suggested, not repeat that detailed analysis here but instead provide a brief summary. The chemical processes that may affect trends of mid-latitude ozone are essentially related to the ODS trends that are known to be responsible for the observed ozone loss in the upper stratosphere and in the polar regions.

Halogen chemistry may lead to ozone depletion in the mid-latitude lower stratosphere through a number of possible mechanisms, including:

- Export of air that has encountered ozone destruction during the winter from the polar vortex.
- Export of air with enhanced levels of active chlorine from the polar vortex.
- In situ activation of chlorine either on cold liquid sulfate aerosol particles or on ice particles. Further, the reaction of N_2O_5 with water on liquid aerosol particles at higher temperatures indirectly enhances the concentrations of ClO at mid-latitudes.
- Ozone depletion due to elevated levels of BrO in the lower stratosphere, possibly caused by transport of very shortlived halogen-containing compounds or BrO across the tropopause.

The first mechanism results from transport combined with polar ozone loss, whereas the remaining three mechanisms all involve in situ chemical destruction of ozone at mid-latitudes. All four mechanisms are ultimately driven by the increase of halogens in the atmosphere over the past decades. Thus, while Millard et al. emphasized the strong interannual variability in the relative contributions of the different mechanisms to the seasonal mid-latitude ozone loss during the winter-spring period, they non-etheless showed that ozone loss driven by catalytic cycles involving halogens was always an important contributor to the simulated mid-latitude ozone loss in the five winters in the 1990s that they studied.

Similarly, Chipperfield found that the observed midlatitude column ozone decrease from 1980 to the early 1990s could be reproduced in long-term simulations in a 3-D chemistry- transport model. The modelled ozone is affected by dynamical interannual variability, but the overall decreases are

Ozone Depletion

dominated by halogen trends; and about 30-50 per cent of the modelled halogen-induced change is a result of high latitude processing on PSCs. Under climate change the strength of the polar vortex may change but the sign of this change is uncertain. A change in the strength and temperature of the vortex will affect chlorine activation and ozone loss there and, through mechanisms and ozone loss in mid-latitudes. The possibility that chlorine might be activated on cirrus clouds or on cold liquid aerosol particles in the lowermost stratosphere was revisited recently by Bregman et al. Based on their model results it seems unlikely that this process is the main mechanism for the observed long-term decline of ozone in the mid-latitude lower stratosphere. Very short-lived organic chlorine-, bromine-, and iodine-containing compounds possess a potential to deplete stratospheric ozone.

However a quantitative assessment of their impact on stratospheric ozone is made difficult by their short lifetime, so there is need to consider the transport pathways from the troposphere to the stratosphere of these compounds in detail. An upper limit for total stratospheric iodine, Iy, of 0.10 ± 0.02 ppt was recently reported by Bösch et al. The impact of this magnitude of iodine loading on stratospheric ozone is negligible. The disturbance of the mid-latitude ozone budget caused by anthropogenic emissions of ODSs will ultimately cease when the stratospheric halogen burden has reached low enough levels. However, like polar ozone, mid-latitude ozone will for many decades remain vulnerable to an enhancement of halogen-catalyzed ozone loss caused by climate change and by natural phenomena such as volcanic eruptions.

IMPACT OF TEMPERATURE CHANGES ON OZONE CHEMISTRY

The increasing abundance of WMGHGs in the atmosphere is expected to lead to an increase in temperature in the troposphere. Furthermore, increasing concentrations of most of these gases, notably CO_2, N_2O and CH_4, are expected to lead to a temperature decrease in the stratosphere. By far the strongest contribution to stratospheric cooling from the WMGHGs comes from CO_2. Temperatures in the stratosphere are also believed to have decreased in part because of the observed reductions in ozone concentrations; any such cooling would have a feedback on the ozone changes. Decreasing stratospheric temperatures lead to a reduction of the ozone loss rate in the upper stratosphere, thereby indirectly leading to more ozone in this region. This reduction of ozone loss is caused by the very strong positive temperature dependence of the ozone loss rate, mainly owing to the Chapman reactions and the NO_x cycle.

This inverse relationship between ozone and temperature changes in the upper stratosphere has been known for many years. More recently, differences in temperature between the two hemispheres have been identified as a cause of inter-hemispheric differences in both the seasonal cycle of upper-stratospheric ozone abundances and in the upper-stratospheric ozone trend deduced from satellite measurements. In the mid-latitude lower stratosphere, one mechanism that leads to ozone loss and is directly sensitive to temperature changes involves heterogeneous reactions on the surfaces of cloud and cold aerosol particles. The rates of many of these reactions increase strongly with decreasing temperature. Similarly, the reaction rates increase with increasing water-vapour concentrations, so that future increases in water, should they occur, would also have an impact.

Note, however, that in the lower-stratospheric mid-latitudes the most important heterogeneous reactions are hydrolysis of N_2O_5 and bromine nitrate, which are relatively insensitive to temperature and water vapour

concentrations. Therefore, the expected impact of climate change on the chemical mechanism is expected to be relatively small. In addition to this impact on heterogeneous chemistry, Zeng and Pyle have argued that a reduction in temperature in the lower stratosphere can slow the rate of HO_x-driven ozone destruction. Polar ozone loss occurs when temperatures in a large enough region sink below the threshold temperature for the existence of PSCs because chlorine is activated by heterogeneous reactions on the surfaces of PSCs. Therefore, a cooling of the stratosphere enhances Arctic ozone loss if the volume of polar air with temperatures below the PSC threshold value increases.

Moreover, a stronger PSC activity is expected to lead to a greater denitrification of the Arctic vortex, hence a slower deactivation of chlorine and, consequently, a greater chemical ozone loss. Rex et al. have deduced an empirical relation between observed temperatures and observed winter-spring chemical loss of Arctic ozone.

Based on this relation, and for current levels of chlorine, about 15 DU additional loss of total ozone is expected for each 1 K cooling of the Arctic lower stratosphere. Although the lower stratosphere is expected to generally cool with increasing greenhouse-gas concentrations, the temperature changes in the lower stratosphere during the polar wintertime will be sensitive to any change in circulation associated with dynamical feedbacks, principally from changes in planetary wave drag. In principle, these feedbacks could be of either sign and thus could lead to enhanced cooling or even to warming.

An early model study found enhanced cooling, to the extent that polar ozone loss would be expected to increase during the next 10 to 15 years even while halogen levels decreased. More recent studies with higher-resolution

Ozone Depletion

models have found a much less dramatic dynamical feedback, with some models showing an increase and some a decrease in planetary wave drag, and with all models predicting a relatively small change in Arctic ozone over the next few decades.

However, it should be noted that model results for the Arctic are difficult to assess because the processes leading to polar ozone depletion show so much natural variability that the atmosphere may evolve anywhere within the envelope provided by an ensemble of model simulations. In any event, it is the Arctic that is most sensitive to the effects caused by climate change. Present- day temperatures in the Arctic lie close to the threshold value for the onset of heterogeneous chlorine activation and thus close to the threshold value for the onset of rapid ozone loss chemistry; in the Antarctic temperatures are much lower and thus ozone loss is not as sensitive to temperature changes.

IMPACT OF METHANE AND WATER VAPOUR CHANGES ON OZONE CHEMISTRY

With the exception of the high latitudes in winter, ozone in the upper stratosphere is in a photochemical steady-state; photochemical reactions are sufficiently fast that ozone concentrations are determined by a local balance between photochemical production and loss. Non-etheless, transport has a significant indirect influence on upper stratospheric ozone insofar as it determines the concentrations of trace compounds such as CH_4, H_2O, CFCs and N_2O, all of which act as precursors of the radicals that determine the ozone chemistry. Furthermore, CH_4 is of particular importance because it is the primary mechanism for the conversion of reactive Cl to the unreactive HCl reservoir via the reaction $CH_4 + Cl' \rightarrow HCl + CH_3$, and thus affects the efficiency of the chlorine-driven ozone loss. Changes in upper stratospheric CH_4 have important implications for upper stratospheric ozone.

Siskind et al. found that for the period 1992–1995, the increase in active chlorine resulting from the CH_4 decrease was the largest contributor to the ozone changes occurring over that time. Indeed, a measured increase in upper- stratospheric ClO between 1991 and 1997, which is significantly greater than that expected from the increase in the chlorine source gases alone, may be explained by the observed concurrent decrease of CH_4.

Further, Li et al. find that inter-hemispheric differences in CH_4 are partly responsible for observed differences in upper stratospheric ozone trends between the two hemispheres. Stratospheric water vapour is the primary source of the HOx radicals that drive the dominant ozone loss cycles in the upper stratosphere. An increase in stratospheric water vapour is therefore expected to lead to a greater chemical loss of ozone in the upper stratosphere. This effect has been investigated quantitatively in model simulations.

In a study of conditions for the year 2010, Jucks and Salawitch find that above 45 km an increase of 1 per cent in the water vapour mixing ratio would completely negate the increase of ozone driven by the 15 per cent decrease of inorganic chlorine that is expected by the year 2010. At 40 km the increase of ozone would still be reduced by about 50 per cent. Shindell conducted a general circulation model study for the period 1979–1996. The simulations show a significant chemical effect of water vapour increases on ozone concentrations, with a reduction of more than 1 per cent between 45 and 55 km, and a maximum impact of about 4 per cent at 50 km. Further, Li et al. found that the annually averaged ozone trend at 45°S and 1.8 hPa increases by 1 per cent per decade for a water vapour increase of 1 per cent yr–1. These model results point to an anticorrelation between ozone and water vapour in the upper stratosphere and lower mesosphere. However, such an anticorrelation is not seen in observations and ozone in this region varies much less than predicted by models.

would also be expected to lead to ozone loss because of an intensified HOx ozone loss cycle. The model results of Dvortsov and Solomon predict that an increase in stratospheric water vapour of 1 per cent yr–1 translates to an additional depletion of mid-latitude column ozone by 0.3 per cent per decade.

In the polar lower stratosphere, the rates of the heterogeneous reactions that are responsible for the activation of chlorine increase with increasing water vapour concentrations. An increase in stratospheric water vapour essentially means that the temperature threshold at which PSCs form, and thus heterogeneous reactions rates begin to become significant for chlorine activation, is shifted to higher values.

If stratospheric water vapour were to increase it could lead to a substantially enhanced Arctic ozone loss in the future. In considering concerning stratospheric water vapour, it needs to be borne in mind that there is considerable uncertainty about the sign of future water vapour changes in light of the puzzling past record.

THE ROLE OF TRANSPORT FOR OZONE CHANGES

Transport is a key factor influencing the seasonal and interannual variability of stratospheric ozone. Seasonal variations in transport force the large winter-spring build-up of extra-tropical total ozone in both hemispheres, and inter-hemispheric differences in transport cause corresponding differences in extra-tropical total ozone. In both hemispheres, mid-latitude ozone decreases during summer and returns to approximate photochemical balance by autumn, and there is a strong persistence of the dynamically forced anomalies throughout summer. The interannual variability in the winter-spring build-up is greater in the NH, reflecting the greater dynamical variability of the NH stratosphere.

Stratospheric Planetary-Wave-induced Transport and Mixing

Large-scale transport of ozone is a result of advection by the Brewer-Dobson circulation and of eddy transport effects; both of these mechanisms are first-order terms in the zonal mean ozone transport equation. The strength of the Brewer-Dobson circulation is directly tied to dissipating planetary waves forced from the troposphere, and eddy transports of ozone are also linked to planetary wave activity, so that net ozone transport is tied to the variability of forced planetary waves. The amount of dissipating wave activity within the stratosphere is related to the vertical component of wave activity entering the lower stratosphere.

This quantity can be derived from conventional meteorological analyses and is a convenient proxy used to quantify PWD. A significant correlation has been found between interannual changes in PWD and total ozone build-up during winter and spring. The fact that the effect of PWD on ozone is seasonal and has essentially no interannual memory suggests that long-term changes in PWD can be expected to lead to long-term changes in winter-spring ozone, all else being equal. However, although the basic physics of the ozone-PWD connection is well understood, its quantification via correlations is at best crude, and this limits our ability to attribute changes in ozone to changes in PWD. In the NH, there have been interannual variations in various meteorological parameters during the period 1980–2000 that together paint a fairly consistent, albeit incomplete, picture. During the mid-1990s the NH exhibited a number of years when the Arctic wintertime vortex was colder and stronger and more persistent.

Any dynamically induced component of these changes requires a weakened Brewer-Dobson circulation, which in turn requires a decrease in PWD. Such a decrease in PWD during this period has been documented, although the results were sensitive to which months and time periods were considered. Randel et al. show that for the period 1979–2000, PWD in the NH increased during early winter and decreased during mid-winter. This seasonal variation is consistent with the Arctic early winter warming and late winter cooling seen over the same time period at 100 hPa. The weakened Brewer-Dobson circulation during midwinter implies a decrease in the winter build-up of mid-latitude ozone, and Randel et al. estimated that the decreased wave driving may account for about 20–30 per cent of the observed changes in ozone in the January to March period. Changes in SH dynamics are not as clear as in the NH, primarily because meteorological reanalysis data sets are less well constrained by observations. Planetary waves, in addition to affecting the temperature and chemistry of the polar stratosphere through the processes described earlier, can also displace the centre of the polar vortex off the pole.

This has important implications for ozone and NOx chemistry because air parcel trajectories within the vortex are then no longer confined to the polar night but experience short periods of sunlight. Polar ozone loss processes are usually limited by the poleward retreat of the terminator. Planetary wave distortion of the vortex can expose deeper vortex air to sunlight and cause ozone depletion chemistry to start earlier than it would have otherwise. In this way, waveinduced displacements of the vortex can drive a mid-winter start to Antarctic ozone depletion. Because wave-induced forcing in the stratosphere is believed to come primarily from planetary-scale Rossby waves that are generated in the troposphere during wintertime, future changes in the generation of tropospheric waves may influence polar ozone abundance. However there is as yet no consensus from models on the sign of this change.

Tropopause Variations and Ozone Mini-Holes

Tropospheric circulation and tropopause height variations also affect the mid-latitude distribution of column ozone. The relationship between local mid- well documented. Day-to-day changes in tropopause height are associated with the passage of synoptic-scale disturbances in the upper troposphere and lower stratosphere, which affect ozone in the lower stratosphere through transport and can result in large local changes in column ozone, particularly in the vicinity of storm tracks. In extreme situations, they can lead to socalled ozone mini-holes, which occur over both hemispheres and have the lowest column ozone levels observed outside the polar vortices, sometimes well under 220 DU. Ozone mini-holes do not primarily entail a destruction of ozone, but rather its re-distribution. Changes in the spatial and temporal occurrence of synoptic-scale processes and Rossby wave breaking that are induced by climate change or natural variability, could lead to changes in the distribution or frequency of the occurrence of mini-hole and low-ozone events, and thus to regional changes in the mean column ozone. Observations have shown that over the NH the altitude of the extra-tropical tropopause has generally increased over recent decades. Radiosonde measurements over both Europe and Canada show an increase in altitude of about 300–600 m over the past 30 years, with the precise amount depending on location. Consistent regional increases are also seen in meteorological reanalyses over both hemispheres, although reanalysis trend studies should be viewed with care. Spatial patterns show increases over the NH and SH beginning at mid-latitudes and reaching a maximum towards the poles.

The magnitude of the changes can also depend on the longitude. However, the effects of tropospheric circulation changes on ozone and the relation of both to tropopause height changes remain poorly understood at present. In particular, the relationship between tropopause height and ozone mentioned earlier applies to single stations; there is no reason to expect it to apply in the zonal mean, or on longer time scales over which ozone transport is irreversible. Because of our poor understanding of what controls the zonal-mean midlatitude tropopause height, or whether it is possible to consider such processes in a zonal-mean approach, it is not clear that the ozone-tropopause height correlations can be extended to decadal time scales in order to estimate changes in ozone.

Some recent model studies suggest that lower stratospheric cooling caused by ozone depletion, acting together with tropospheric warming due to WMGHGs, can be a contributor to tropopause height changes in which case the tropopause height changes cannot be entirely considered as a cause of the ozone changes. One source of uncertainty in these model calculations is that most climate models cannot resolve the observed tropopause height changes, so these must be inferred by interpolation; another is that the observed ozone changes are not well quantified close to the tropopause. Furthermore there remain considerable differences between the various reanalysis products available, as well as between the model results.

Stratosphere-Troposphere Exchange

Stratosphere-troposphere exchange processes also affect the ozone distribution and have the capability to affect tropospheric chemistry. STE is a two-way process that encompasses transport from the troposphere into the stratosphere and from the stratosphere into the troposphere through a variety of processes. Some aspects of STE are implicit in the preceding discussions. The net mass flux from the troposphere into the stratosphere and back again is driven by PWD through the Brewer-Dobson circulation. Although the global approach can explain the net global flux, local and regional processes need to be identified and assessed to fully understand the distribution of trace species being exchanged. Such understanding is relevant both for chemical budgets and for climate change and variability studies. Although there is a net TST in the tropics and a net STT in the extratropics, a number of mechanisms that lead to STT in the tropics and TST in the extratropics have been identified. In the extratropics TST can significantly affect the composition of the lowermost stratosphere, even if the exchange cannot reach higher into the stratosphere, by introducing, during deep TST events, near-surface pollutants, and ozone- poor and humid air.

Similarly, deep STT events can transport ozone-rich air into the mid- and lower troposphere. It should be noted, however, that so-called shallow events, with exchanges near the tropopause, remain the main feature in the extratropics. Regions of occurrence of such exchange processes, at least in the NH, appear to be linked to the storm tracks and their variability, that is, in the same region where mini-holes are most frequent. The past and future variability of STE remains to be assessed. It is clear that changes in planetary wave activity will modify the Brewer-Dobson circulation and hence affect global transport processes. As for regional mechanisms, seasonal and interannual variability, driven, for example, by the North Atlantic Oscillation and the El Niño-Southern Oscillation have already been observed. Changes in the occurrence of regional and local weather phenomena could also modify STE. Fifteen-year studies with reanalysis products have not yielded any distinct trends. Given the discrepancies between different approaches to evaluate STE and inhomogeneities in meteorological analyses and reanalyses, consistent trend studies are not available as of yet.

The Tropical Tropopause Layer

Air enters the stratosphere primarily in the tropics, and hence the physical and chemical characteristics of air near the tropical tropopause behave as boundary conditions for the global stratosphere. The tropical tropopause is relatively high, near 17 km. The tropospheric lapse rate is determined by radiative-convective equilibrium, whereas the thermal structure above 14 km is primarily in radiative balance, which is characteristic of the stratosphere. Overall, the region of the tropical atmosphere between about 12 km and the tropopause has characteristics intermediate to those of the troposphere and stratosphere, and is referred to as the tropical tropopause layer. Thin cirrus clouds are observed over large areas of the TTL, although their formation mechanism(s) and effects on large-scale circulation are poorly known. In the tropics, the background clear-sky radiative balance shifts from cooling in the troposphere to heating in the stratosphere, with the transition occurring at around 15 km. The region of heating above about 15 km is linked to mean upward motion into the lower stratosphere, and this region marks the base of the stratospheric Brewer-Dobson circulation.

The TTL is coupled to stratospheric ozone through its control of stratospheric water vapour and by the transport of tropospheric source gases and ODSs into the lower stratosphere, so changes in the TTL could affect

Ozone Depletion

the stratospheric ozone layer. If the residence time of air in the TTL before it is transported into the stratosphere is short, then it may be possible that short-lived halogen compounds or their degradation products could enter the lower stratosphere and contribute to ozone loss there. Air entering the stratosphere is dehydrated as it passes through the cold tropical tropopause, and this drying accounts for the extreme aridity of the global stratosphere. Furthermore, the seasonal cycle in tropopause temperature imparts a strong seasonal variation in stratospheric water vapour, which then propagates with the mean stratospheric transport circulation.

Year-to-year variations in tropical tropopause temperatures are also highly correlated with global stratospheric water vapour anomalies, although there remain some issues concerning the consistency of decadal-scale changes in water vapour. However, while there is strong empirical coupling between tropopause temperatures and stratospheric water vapour, details of the dehydration mechanism(s) within the TTL are still a topic of scientific debate. One critical, unanswered question is whether, and how, the TTL will change in response to climate change and what will be the resulting impact on stratospheric ozone.

STRATOSPHERE-TROPOSPHERE DYNAMICAL COUPLING

Analysis of observational data shows that atmospheric circulation tends to maintain spatially coherent, large-scale patterns for extended periods of time, and then to shift to similar patterns of opposite phase. These patterns represent preferred modes of variability of the coupled atmosphere-ocean-land-sea-ice system. They fluctuate on intra-seasonal, seasonal, interannual and decadal time scales, and are influenced by externally and anthropogenically caused climate variability. Some of the patterns exhibit seesaw-like behaviour and are usually called oscillations. A few circulation modes have been implicated in stratosphere-troposphere dynamical coupling, and thus may provide a coupling between stratospheric ozone depletion and tropospheric climate.

- *Northern Annular Mode*: The NAM, also referred to as the Arctic Oscillation is a hemispherewide annular atmospheric circulation pattern in which atmospheric pressure over the northern polar region varies out of phase with pressure over northern mid-latitudes, on time scales ranging from weeks to decades.
- *Southern Annular Mode*: The SAM, also referred to as the Antarctic Oscillation, is the SH analogue of the NAM. The SAM exhibits a large-scale alternation of pressure and temperature between the mid-latitudes and the polar region.

- *North Atlantic Oscillation*: The NAO was originally identified as a seesaw of sea-level pressure between the Icelandic Low and the Azores High, but an associated circulation pattern is also exhibited well above in the troposphere. The NAO is the dominant regional pattern of wintertime atmospheric circulation variability over the extra-tropical North Atlantic, and has exhibited variability and trends over long time periods. The relationship between the NAO and the NAM remains a matter of debate.

The NAM extends through the depth of the troposphere. During the cold season when the stratosphere has large-amplitude disturbances, the NAM also has a strong signature in the stratosphere, where it is associated with variations in the strength of the westerly vortex that encircles the Arctic polar stratosphere; this signature suggests a coupling between the stratosphere and the troposphere. During winters when the stratospheric vortex is stronger than normal, the NAM tends to be in a positive phase. Circulation modes can affect the ozone distribution directly and indirectly by influencing propagation of planetary waves from the troposphere into the middle atmosphere. Therefore, changes in circulation modes can produce changes in ozone distribution.

It has been shown that because the NAO has a large vertical extent during the winter and because it modulates the tropopause height, it can explain much of the spatial pattern in column ozone trends in the North Atlantic over the past 30 years. Stratospheric changes may feed back onto changes in circulation modes. Observations suggest that at least in some cases the large amplitude NAM anomalies tend to propagate from the stratosphere to the troposphere on time scales of weeks to a few months.

Furthermore, because of the strong coupling between the stratospheric vortex and the NAM, the recent trend in the NAM and NAO has been associated with processes that are known to affect the strength of the stratospheric polar vortex, such as tropical volcanic eruptions, ozone depletion and anthropogenic changes in greenhouse gas concentrations. However, some modelling studies have shown that a simulated trend in the tropospheric NAM and SAM does not necessarily depend on stratospheric involvement. Global climate modelling simulations that include interactive stratospheric chemistry suggest that one mechanism by which solar variability may affect tropospheric climate is through solar-forced changes in upper-stratospheric ozone that induce changes in the leading mode of variability of the coupled troposphere- stratosphere circulation. A number of modelling studies have examined the effect of increased concentrations of greenhouse gases on the annular modes.

Most coupled atmosphereocean climate models agree in finding a positive NAM trend under increasing greenhouse-gas concentrations, a trend which is qualitatively consistent with the observed positive NAM trend. A more positive NAM is consistent with a stronger stratospheric vortex. On the other hand, an intensified polar vortex is related to changes in planetary- and synoptic-scale wave characteristics, and may produce tropospheric circulation anomalies similar to the positive phase of the NAM. These potential feedbacks obscure cause and effect. For example, whereas Rind et al. and Sigmond et al. found that in the middle stratosphere the perturbation to the NAM from a CO_2 climate depends on modelled changes in sea-surface temperatures, Sigmond et al. suggested that perturbations in the zonal wind near the surface are mainly caused by doubling of stratospheric CO_2.

The stratosphere-troposphere dynamical coupling through the circulation modes discussed in this part suggests that stratospheric dynamics should be accounted for in the problem of detection and prediction of future tropospheric climate change. Because increased concentrations of greenhouse gases may cause changes in stratospheric dynamics, greenhouse- gas changes may induce changes in surface climate through stratosphere-troposphere dynamical coupling, in addition to radiative forcing.

POSSIBLE DYNAMICAL FEEDBACKS OF OZONE CHANGES

The effect of ozone loss on polar stratospheric temperatures results in concomitant changes in stratospheric zonal winds and polar vortex structure, and also possible changes in planetarywave behaviour. The strongest effect is seen in the Antarctic, where a large cooling of the lower stratosphere, which is associated with the ozone hole has resulted in an intensified and more persistent springtime polar vortex. Waugh et al. show that since the ozone hole developed, the break-up date of the Antarctic vortex occurs two to three weeks later than before. Recent research has shown that trends in surface temperatures over Antarctica may be in part traceable, at least over the period 1980–2000, to trends in the lower stratospheric polar vortex, which are largely caused by the ozone hole.

It has been suggested that during the early summer the strengthening of the westerly flow extends all the way to the surface. The role of this mechanism– which has yet to be elucidated–has been highlighted by the modelling study of Gillett and Thompson who prescribed ozone depletion in an atmospheric climate model. They found that the seasonality, structure and amplitude of the modelled changes in 500 hPa geopotential height and near-surface air temperature in the Antarctic had similar spatial patterns to observations, which were a cooling over the Antarctic interior and a warming over the peninsula and South America. This result suggests that anthropogenic emissions of ozone-depleting gases have had a distinct impact on climate not only in the stratosphere but also at the Earth's surface. These surface changes appear to act in the same direction as changes resulting from increases in greenhouse gases.

An increase in the strength of the Antarctic polar vortex by stratospheric ozone depletion does not affect only surface winds and temperatures. Using a 15,000-year integration of a coupled ocean-atmosphere model, Hall and Visbeck showed that fluctuations of the mid-latitude westerly winds generate ocean circulation and sea-ice variations on interannual to centennial time scales. Cause and effect are more difficult to separate in the Arctic stratosphere than in the Antarctic stratosphere because of higher 'natural' meteorological variability, relatively smaller ozone losses and a less clear separation between transport and chemical effects on ozone. Furthermore, temperatures need to be sufficiently low in any given year to initiate polar ozone chemistry.

Non-etheless, in several years during the mid-1990s the Arctic experienced low ozone, low temperatures in late spring and enhanced vortex persistence which resembled the coupled changes observed in the Antarctic. A crucial issue is the potential feedback between changes in stratospheric climate and planetary wave forcing. However, this topic is poorly understood at present.

Chen and Robinson and Limpasuvan and Hartmann have argued that stronger vertical shear of the zonal wind at high latitudes will reduce the strength of stratospheric PWD, whereas Hu and Tung have argued for precisely the opposite effect.

Many climate model simulations found enhanced PWD in an atmosphere with elevated WMGHG concentrations, which leads to a stronger Brewer-Dobson circulation; however models were not unanimous on this result.

Because planetary wave forcing is a primary driver of the stratospheric circulation, improved understanding of this coupling will be necessary to predict future ozone changes.

OBSERVED CHANGES IN STRATOSPHERIC OZONE

Global and hemispheric-scale variations in stratospheric ozone can be quantified from extensive observational records covering the past 20 to 30 years. There are numerous ways to measure ozone in the atmosphere, but they fall broadly into two categories: measurements of column ozone and measurements of the vertical profile of ozone. Approximately 90 per cent of the vertically integrated ozone column resides in the stratosphere. There are more independent measurements, longer time-series, and better global coverage for column ozone. Regular measurements of column ozone are available from a network of surface stations, mostly in the mid-latitude NH, with reasonable coverage extending back to the 1960s. Near-global, continuous column ozone data are available from satellite measurements beginning in 1979. The different observational data sets can be used to estimate past ozone changes, and the differences between data sets provide a lower bound of overall uncertainty.

The differences indicate good overall agreement between different data sources for changes in column ozone, and thus we have reasonable confidence in describing the spatial and temporal characteristics of past changes. Five data sets of zonal and monthly mean column ozone values developed by different scientific teams were used to quantify past ozone changes; they include ground-based measurements covering 1964–2001, and several different satellite data sets extending in time over 1979–2001. The analyses first remove the seasonal cycle from each data set and the deviations are area weighted and expressed as anomalies with respect to the period 1964–1980. The global ozone amount shows decreasing values between the late 1970s and the early 1990s, a relative minimum during 1992–1994, and slightly increasing values during the late 1990s.

Global ozone for the period 1997–2004 was approximately 3 per cent below the 1964–1980 average values. Since systematic global observations began in the mid-1960s, the lowest annually averaged global ozone occurred during 1992–1993. These changes are evident in each of the available global data sets. No significant long-term changes in column ozone have been observed in the tropics. Column ozone changes averaged over mid-latitudes are significantly larger in the SH than in the NH; averaged for the period 1997–2001, SH values are 6 per cent below pre-1980 values, whereas NH values are 3 per cent lower. Also, there is significant seasonality to the NH mid-latitude losses whereas long-term SH losses are about 6 per cent year round. The most dramatic changes in ozone have occurred during the spring season over Antarctica, with the development during the 1980s of a phenomenon known as the ozone hole.

The ozone hole now recurs every spring, with some interannual variability and occasional extreme behaviour. In most years the ozone concentration is reduced to nearly zero over a layer several kilometres deep within the lower stratosphere in the Antarctic polar vortex. Since the early 1990s, the average October column ozone poleward of 63°S has been more than 100 DU below pre-ozone-hole values with up to a 70 per cent local decrease for periods of a week or so. Compared with the Antarctic, Arctic ozone abundance in the winter and spring is highly variable because of interannual variability in chemical loss and in dynamical transport. Dynamical variability within the winter stratosphere leads to changes in ozone transport to high latitudes, and these transport changes are correlated with polar temperature variability– with less ozone transport being associated with lower temperatures. Low temperatures favour halogen-induced chemical ozone loss. Thus, in recent decades, halogen- induced polar ozone chemistry has acted in concert with the dynamically induced ozone variability, and has led to Arctic column ozone losses of up to 30 per cent in particularly cold winters. In particularly dynamically active, warm winters, however, the estimated chemical ozone loss has been very small. Changes in the vertical profile of ozone are derived primarily from satellites, ground-based measurements and balloon-borne ozonesondes. Long records from ground-based and balloon data are available mainly for stations over NH mid-latitudes. Ozone profile changes over NH mid-latitudes exhibit two maxima, with decreasing trends in the upper stratosphere and in the lower stratosphere during the period 1979–2000. Ozone profile trends show a minimum near 30 km. The vertically integrated profile trends are in agreement with the measured changes in column ozone.

OBSERVED CHANGES IN ODSS

As a result of reduced emissions because of the Montreal Protocol and its Amendments and Adjustments, mixing ratios for most ODSs have stopped increasing near the Earth's surface. The response to the Protocol, however, is reflected in quite different observed behaviour for different substances. By 2003, the mixing ratios for CFC-12 were close to their peak, CFC-11 had clearly decreased, while methyl chloroform had dropped by 80 per cent from its maximum. Halons and HCFCs are among the few ODSs whose mixing ratios were still increasing in 2000. Halons contain bromine, which is on average 40 to 50 times more efficient on a per-atom basis at destroying stratospheric ozone than chlorine.

However, growth rates for most halons have steadily decreased during recent years. Furthermore, increases in tropospheric bromine from halons have been offset by the decline observed for methyl bromide since 1998. Atmospheric amounts of HCFCs continue to increase because of their use as CFC substitutes. Chlorine from HCFCs has increased at a fairly constant rate of 10 ppt Cl yr–1 since 1996, although HCFCs accounted for only 5 per cent of all chlorine from long-lived gases in the atmosphere by 2000. The ODPs of the most abundant HCFCs are only about 5–10 per cent of those of the CFCs. The behaviour of CFC-11 is representative of the behaviour of CFCs in general, and the behaviour of HCFC-22 and HFC-134a is representative of HCFCs in general. The most rapid growth rates of CFC-11 occurred in the 1970s and 1980s. The largest emissions were in the NH, and concentrations in the SH lagged behind those in the NH, consistent with an inter-hemispheric mixing time scale of about 1 to 2 years.

In recent years, following the implementation of the Montreal Protocol, the observed growth rate has declined, concentrations appear to be at their peak and, as emissions have declined, the inter-hemispheric gradient has almost disappeared. In contrast, HCFC-22 and HFC-134a concentrations are still growing rapidly and there is a marked inter-hemispheric gradient. Ground-based observations suggest that by 2003 the cumulative totals of both chlorine- and bromine-containing gases regulated by the Montreal Protocol were decreasing in the lower atmosphere. Although tropospheric chlorine levels peaked in the early 1990s and have since declined, atmospheric bromine began decreasing in 1998. The net effect of changes in the abundance of both chlorine and bromine on the total ozone-depleting halogens in the stratosphere is estimated roughly by calculating the equivalent effective stratospheric chlorine. The calculation of EESC includes consideration of the total amount of chlorine and bromine accounted for by long-lived halocarbons, how rapidly these halocarbons degrade and release their halogen in the stratosphere and a nominal lag time of three years to allow for transport from the troposphere into the stratosphere. The tropospheric observational data suggest that EESC peaked in the mid-1990s and has been decreasing at a mean rate of 22 ppt yr–1 over the past eight years. Direct stratospheric measurements show that stratospheric chlorine reached a broad plateau after 1996, characterized by variability.

OBSERVED CHANGES IN STRATOSPHERIC AEROSOLS, WATER VAPOUR, METHANE AND NITROUS OXIDE

In addition to ODSs, stratospheric ozone is influenced by the abundance of stratospheric aerosols, water vapour, methane and nitrous oxide. During the past three decades, aerosol loading in the stratosphere has primarily reflected the effects of a few volcanic eruptions that inject aerosol and its gaseous precursors into the stratosphere. The most noteworthy of these eruptions are El Chichón and Pinatubo. The 1991 Pinatubo eruption likely had the largest impact of any event in the 20th century, producing about 30 Tg of aerosol that persisted into at least the late 1990s. Current aerosol loading, which is at the lowest observed levels, is less than 0.5 Tg, so the Pinatubo event represents nearly a factor of 100 enhancement relative to non-volcanic levels. The source of the non-volcanic stratospheric aerosol is primarily carbonyl sulfide and there is general agreement between the aerosols estimated by modelling the transformation of observed OCS to sulphate aerosols and observed aerosols.

However, there is a significant dearth of SO2 measurements, and the role of tropospheric SO2 in the stratospheric aerosol budget–while significant–remains a matter of some uncertainty. Because of the high variability of

stratospheric aerosol loading it is difficult to detect trends in the non-volcanic aerosol component. Trends derived from the late 1970s to the current period are likely to encompass a value of zero. The recent Stratospheric Processes And their Role in Climate Assessment of Upper Tropospheric and Stratospheric Water Vapour provided an extensive review of data sources and quality for stratospheric water vapour, together with detailed analyses of observed seasonal and interannual variability. The longest continuous reliable data set is at a single location is based on balloon-borne frost-point hygrometer measurements and dates back to 1980. Over the period 1980–2000, a statistically significant positive trend of approximately 1 per cent yr–1 is observed at all levels between about 15 and 26 km in altitude. However, although a linear trend can be fitted to these data, there is a high degree of variability in the infrequent sampling, and the increases are neither continuous nor steady. Long- term increases in stratospheric water vapour are also inferred from a number of other ground-based, balloon, aircraft and satellite data sets spanning approximately 1980–2000, although the time records are short and the sampling uncertainty is high in many cases.

Global stratospheric water vapour measurements have been made by the Halogen Occultation Experiment satellite instrument for more than a decade. Interannual changes in water vapour derived from HALOE data show excellent agreement with Polar Ozone and Aerosol Measurement satellite data, and also exhibit strong coherence with tropical tropopause temperature changes. The Boulder and HALOE data show reasonable agreement for the early part of the record but there is an offset after 1997, with the Boulder data showing higher values than HALOE measurements. As a result, linear trends derived from these two data sets for the period 1992–2004 show very different results. The reason for the differences between the balloon and satellite data is unclear at present, but the discrepancy calls into question interpretation of water vapour trends derived from short or infrequently sampled data records.

It will be important to reconcile these differences because these data sets are the two longest and most continuous data records available for stratospheric water vapour. It is a challenge to explain the magnitude of the water vapour increases seen in the Boulder frost-point data. Somewhat less than half the observed increase of about 10 per cent per decade can be explained as a result of increasing tropospheric methane. The remaining increase could be reconciled with a warming of the tropical tropopause of approximately 1 K per decade, assuming that air entered the stratosphere

Ozone Depletion

with water vapour in equilibrium with ice. However, observations suggest that the tropical tropopause has cooled slightly for this time period, by approximately 0.5 K per decade, and risen slightly in altitude by about 20 m per decade. Although regional-scale processes may also influence stratospheric water vapour, there is no evidence for increases in tropopause temperature in these regions either. From this perspective, the extent of the decadal water vapour increases inferred from the Boulder measurements is inconsistent with the observed tropical tropopause cooling, and this inconsistency limits confidence in predicting the future evolution of stratospheric water vapour. The atmospheric abundance of methane has increased by a factor of about 2.5 since the pre-industrial era. Measurements of methane from a global monitoring network showed increasing values through the 1990s, but approximately constant values during 1999–2002. Changes in stratospheric methane have been monitored on a global scale using HALOE satellite measurements since 1991. The HALOE data show increases in lower stratospheric methane during 1992–1997 that are in reasonable agreement with tropospheric increases during this time.

In the upper stratosphere the HALOE data show an overall decrease in methane since 1991, which is likely attributable to a combination of chemical and dynamical influences. Measurements of tropospheric N2O show a consistent increase of about 3 per cent per decade. Because tropospheric air is transported into the stratosphere, these positive N2O trends produce increases in stratospheric reactive nitrogen which plays a key role in ozone photochemistry. Measurements of stratospheric column NO2 from the SH and the NH mid-latitudes show long-term increases of approximately 6 per cent per decade and these are reconciled with the N2O changes by considering effects of changing levels of stratospheric ozone, water vapour and halogens.

OBSERVED TEMPERATURE CHANGES IN THE STRATOSPHERE

There is strong evidence of a large and significant cooling in most of the stratosphere since 1980. Recent updates to the observed changes in stratospheric temperature and to the understanding of those changes have been presented of WMO. Current long-term monitoring of stratospheric temperature relies on satellite instruments and radiosonde analyses. The Microwave Sounding Unit and Stratospheric Sounding Unit instruments record temperatures in several 10–15 km thick layers between 17 and 50 km in altitude. Radiosonde trend analyses are available up to altitudes of roughly 25 km. Determining accurate trends with these data sets is difficult. In particular the radiosonde coverage is not global and suffers from data quality concerns, whereas the satellite trend data is a result of merging data sets from several different instruments. It reveals a strong imprint of 1 to 2 years of warming following the volcanic eruptions of Agung, El Chichón and Mt. Pinatubo. When these years are excluded from long-term trend analyses, a significant global cooling is seen in both the radiosonde and the satellite records over the last few decades. This cooling is significant at all levels of the stratosphere except the 30 km level in the SSU record. The largest global cooling is found in the upper stratosphere, where it is fairly uniform in time at a rate of about 2 K per decade. In the lower stratosphere this long-term global cooling manifests itself as more of a step-like change following the volcanic warming events. The cooling also varies with latitude.

The lower stratosphere extratropics show a cooling of 0.4–0.8 K per decade in both hemispheres, which remains roughly constant throughout the year. At high latitudes most of the cooling, up to 2 K per decade for both hemispheres, occurs during spring. Much recent progress has been made in modelling these temperature trends. Models range from one-dimensional fixed dynamical heating rate calculations to three-dimensional coupled chemistry- climate models; many of their findings are presented later in this stage. FDH is a simple way of determining the radiative response to an imposed change whilst 'fixing' the background dynamical heating to its climatological value; this makes the calculation of temperature change simpler, as only a radiation model needs to be used. Changes in dynamical circulations simulated by models can lead to effects over latitudinal bands, which vary between models.

However, dynamics cannot easily produce a global mean temperature change. Because of this, global mean temperature is radiatively controlled and provides an important focus for attribution. For the global mean in the upper stratosphere the models suggest roughly equal contributions to the cooling from ozone decreases and carbon dioxide increases. In the global mean mid-stratosphere there appears to be some discrepancy between the SSU and modelled trends near 30 km: models predict a definite radiative cooling from carbon dioxide at these altitudes, which is not evident in the SSU record. In the lower stratosphere the cooling from carbon dioxide is much smaller than higher in the stratosphere. Although ozone depletion probably accounts for up to half of the observed cooling trend, it appears that a significant cooling from another mechanism may be needed to account for the rest of the observed cooling.

One possible cause of this extra cooling could be stratospheric water vapour increases. However, stratospheric water vapour changes are currently too uncertain to pinpoint their precise role. Tropospheric ozone increases have probably contributed slightly to lower stratospheric cooling by reducing

Ozone Depletion

upwelling thermal radiation; one estimate suggests a cooling of 0.05 K per decade at 50 hPa and a total cooling of up to 0.5 K over the last century. The springtime cooling in the Antarctic lower stratosphere is almost certainly nearly all caused by stratospheric ozone depletion. However, the similar magnitude of cooling in the Arctic spring does not seem to be solely caused by ozone changes, which are much smaller than in the Antarctic; interannual variability may contribute substantially to the cooling observed in the Arctic wintertime. One mechanism for altering temperatures in the upper troposphere and lower stratosphere comes from the direct radiative effects of halocarbons. In contrast with the role of carbon dioxide, halocarbons can actually warm the upper tropospheric and lower stratospheric region.

In summary, stratospheric temperature changes over the past few decades are significant and there are clear quantifiable features of the contributions from ozone, carbon dioxide and volcanism in the past stratospheric temperature record. More definite attribution of the causes of these trends is limited by the short timeseries. Future increases of carbon dioxide can be expected to substantially cool the upper stratosphere.

However, this cooling could be partially offset by any future ozone increase. In the lower stratosphere, both ozone recovery and halocarbon increases would warm this region compared with the present. Any changes in stratospheric water vapour or changes in tropospheric conditions, such as high-cloud properties and tropospheric ozone, would also affect future temperatures in the lower stratosphere. Furthermore, circulation changes can affect temperatures over sub-global scales, especially at midand polar latitudes. Several studies have modelled parts of these expected temperature changes.

OZONE IN THE ATMOSPHERE AND ITS ROLE IN CLIMATE

A number of important concepts concerning the ozone layer and its role in the stratosphere. Ozone, like water vapour and carbon dioxide, is an important and naturally occurring greenhouse gas; that is, it absorbs and emits radiation in the thermal infrared, trapping heat to warm the Earth's surface. In contrast to the so-called wellmixed greenhouse gases, stratospheric ozone has two distinguishing properties. First, its relatively short chemical lifetime means that it is not uniformly mixed throughout the atmosphere and therefore its distribution is controlled by both dynamical and chemical processes. In fact, unlike the WMGHGs, ozone is produced entirely within the atmosphere rather than being emitted into it. Second, it is a very strong absorber of short wavelength UV radiation.

The ozone layer's absorption of this UV radiation leads to the characteristic increase of temperature with altitude in the stratosphere and, in consequence, to a strong resistance to vertical motion. As well as ozone's role in climate it also has more direct links to humans: its absorption of UV radiation protects much of Earth's biota from this potentially damaging short wavelength radiation. In contrast to the benefits of stratospheric ozone, high surface ozone values are detrimental to human health. The distribution of ozone in the atmosphere is maintained by a balance between photochemical production and loss, and by transport between regions of net production and net loss. A number of different chemical regimes can be identified for ozone. In the upper stratosphere, the ozone distribution arises from a balance between production following photolysis of molecular oxygen and destruction via a number of catalytic cycles involving hydrogen, nitrogen and halogen radical species. The halogens arise mainly from anthropogenic ODSs. In the upper stratosphere, the rates of ozone destruction depend on temperature and on the concentrations of the radical species. In the lower stratosphere, reactions on aerosols become important. The distribution of the radicals can be affected by heterogeneous and multiphase chemistry acting on condensed matter. At the low temperature of the wintertime polar lower stratosphere, this is the chemistry that leads to the ozone hole.

The large-scale circulation of the stratosphere, known as the Brewer-Dobson circulation, systematically transports ozone poleward and downward. Because ozone photochemical reactions proceed quickly in the sunlit upper stratosphere, this transport has little effect on the ozone distribution there as ozone removal by transport is quickly replenished by photochemical production. However, this transport leads to significant variations of ozone in the extra-tropical lower stratosphere, where the photochemical relaxation time is very long and ozone can accumulate on seasonal time scales. Due to the seasonality of the Brewer- Dobson circulation, ozone builds up in the extra- tropical lower stratosphere during winter and spring through transport, and then decays photochemically during the summer when transport is weaker. The columnozone distribution is dominated by its distribution in the lower stratosphere and reflects this seasonality. Furthermore, planetary waves are stronger in the Northern Hemisphere than in the Southern Hemisphere, because of the asymmetric distribution of the surface features that, in combination with surface winds, force the waves.

The stratospheric Brewer- Dobson circulation is stronger during the NH winter, and the resulting extra-tropical build-up of ozone during the winter and spring is greater in the NH than in the SH. Variations in the Brewer- Dobson circulation also influence polar temperatures in the lower stratosphere; stronger wave forcing coincides with enhanced circulation and higher polar temperatures. Since temperature affects ozone chemistry, dynamical and chemical effects on column ozone thus tend to act in concert and are coupled. Human activities have led to changes in the atmospheric concentrations of several greenhouse gases, including tropospheric and stratospheric ozone and ODSs and their substitutes. Changes to the concentrations of these gases alter the radiative balance of the Earth's atmosphere by changing the balance between incoming solar radiation and outgoing infrared radiation. Such an alteration in the Earth's radiative balance is called a radiative forcing. This report, past IPCC reports and climate change protocols have universally adopted the concept of radiative forcing as a tool to gauge and contrast surface climate change caused by different mechanisms. Positive radiative forcings are expected to warm the Earth's surface and negative radiative forcings are expected to cool it. Changes in carbon dioxide provide the largest radiative forcing term and are expected to be the largest overall contributor to climate change. In contrast with the positive radiative forcings due to increases in other greenhouse gases, the radiative forcing due to stratospheric ozone depletion is negative. Halocarbons are particularly effective greenhouse gases in part because they absorb the Earth's outgoing infrared radiation in a spectral range where energy is not removed by carbon dioxide or water vapour.

Halocarbon molecules can be many thousands of times more efficient at absorbing the radiant energy emitted from the Earth than a molecule of carbon dioxide, which explains why relatively small amounts of these gases can contribute significantly to radiative forcing of the climate system. Because halocarbons have low concentrations and absorb in the atmospheric window, the magnitude of the direct radiative forcing from a halocarbon is given by the product of its tropospheric mixing ratio and its radiative efficiency. In contrast, for the more abundant greenhouse gases there is a non-linear relationship between the mixing ratio and the radiative forcing. Since 1970 the growth in halocarbon concentrations and the changes in ozone concentrations have been very significant contributors to the total radiative forcing of the Earth's atmosphere.

Because halocarbons have likely caused most of the stratospheric ozone loss, there is the possibility of a partial offset between the positive forcing of halocarbon that are ODSs and the negative forcing from stratospheric ozone loss. The climate impacts of ozone changes are not confined to the surface: stratospheric ozone changes are probably responsible for a significant fraction of the observed cooling in the lower stratosphere over the last two decades and may alter atmospheric dynamics and chemistry. Further, it was predicted that depletion of stratospheric ozone would lead to a global increase in erythemal UV radiation at the surface of about 3 per cent, with much larger increases at high latitudes; these predicted high latitude increases in surface UV dose have indeed been observed.

CHAPTER OUTLINE

Our aim in this stage is to review the scientific understanding of the interactions between ozone and climate. We interpret the term 'climate' broadly, to include stratospheric temperature and circulation, and refer to tropospheric effects as 'tropospheric climate'. This is a broad topic; our review will take a more restricted view concentrating on the interactions as they relate to stratospheric ozone and the role of the ODSs and their substitutes. For this reason the role of tropospheric ozone in the climate system is mentioned only briefly. Similarly, some broader issues, including possible changes in stratospheric water vapour and climate-dependent changes in biogenic emissions will not be discussed in any detail. The relationship between ozone and the solar cycle is well established, and the solar cycle has been used as an explanatory variable in ozone trend analysis. An update on stratospheric observations of ozone, ozone-related species and temperature. A process level, the various feedbacks connecting stratospheric ozone and the climate system. The understanding of these processes leads to a discussion of the attribution of past changes in ozone, the prediction of future changes, and the connection of both with climate. Finally, reviews the trade-offs between ozone depletion and radiative forcing, focusing on ODSs and their substitutes.

OZONE DEPLETING SUBSTANCES

CFCs are not the only ozone-depleting chemicals (ODCs). A number of other halocarbon species are capable of destroying ozone in the stratosphere. Halocarbons include the chlorofluorocarbons (CFCs), the hydrochlorofluorocarbons (HCFCs), methylhalides, carbon tetrachloride (CCl_4), carbon tetrafluoride (CF_4), and the halons (bromide species). Although the HCFCs do not contribute significantly to the destruction of the ozone layer, they, along with the other halocarbons are all considered to be powerful greenhouse gases and contribute towards global warming.

HYDROCHLOROFLUOROCARBONS

Hydrochlorofluorocarbons or HCFCs contain chlorine but, unlike CFCs, they also contain hydrogen which causes them to break down in the lower atmosphere (troposphere). They are called transition chemicals because they are considered an interim step between strong ozone-depleters and replacement chemicals that are entirely ozone-friendly. Unfortunately, like CFCs, they are strong greenhouse gases and contribute towards global warming.

CARBON TETRACHLORIDE

Carbon tetrachloride (CCl4), despite its toxicity, was first used in the early 1900s as a fire extinguishant, and more recently as an industrial solvent, an agricultural fumigant, and in many other industrial processes including petrochemical refining, and pesticide and pharmaceuticals production.

Recently it has also been used in the production of CFC-11 and CFC-12. It has accounted for less than 8 per cent of total ozone depletion. The use of carbon tetrachloride in developed countries has been prohibited since the beginning of 1996 under the Montreal Protocol.

METHYL CHLOROFORM

Methyl chloroform, also known as 1,1,1 trichloroethane is a versatile, all-purpose industrial solvent used primarily to clean metal and electronic parts. It was introduced in the 1950s as a substitute for carbon tetrachloride. Methyl chloroform has accounted for roughly 5 per cent of total ozone depletion. The use of methyl chloroform in developed countries has been prohibited since the beginning of 1996 under the Montreal Protocol.

HALONS

Halons, unlike CFCs, contain bromine, which also destroys ozone in the stratosphere. Halons are used primarily as fire suppressants. Halon-1301 has an ozone depleting potential 10 times that of CFC-11. Although the use of halons in developed countries has been phased out since 1996, the atmospheric concentration of these potent, long-lived ozone destroyers is still rising by an estimated 11 to 15 per cent annually. To date halons have accounted for about 5 per cent of global ozone depletion.

METHYL BROMIDE

Methyl bromide, another bromine-containing halocarbon, has been used as a pesticide since the 1960s. Today, scientists estimate that human sources of methyl bromide are responsible for approximately 5 to 10 per cent of global ozone depletion.

THE OZONE HOLE: ANTARCTICA

Why is the ozone hole over Antarctica? That is one of the first questions that comes to mind when people think about the ozone hole. Every winter and spring since the late 1970s, an ozone hole has formed in the stratosphere above the Antarctic continent. In recent years this hole has become both larger and deeper, in the sense that more and more ozone is being destroyed. As summer approaches, the hole repairs itself, only to reform during the following spring.

MEASURING THE OZONE HOLE

The most common ozone measurement unit is the Dobson Unit (DU). The Dobson Unit is named after atmospheric ozone pioneer G.M.B. Dobson who carried out the earliest studies on ozone in the atmosphere from the 1920s to the 1970s. A DU measures the total amount of ozone in an overhead column of the atmosphere. Dobson Units are measured by how thick the layer of ozone would be if it were compressed into one layer at 0 degrees Celsius and with a pressure of one atmosphere above it. Every 0.01 millimetre thickness of the layer is equal to one Dobson Unit. The average amount of ozone in the stratosphere across the globe is about 300 DU (or a thickness of only 3mm at 0^oC and 1 atmospheric pressure!). Highest levels of ozone are usually found in the mid to high latitudes, in Canada and Siberia (360DU).

WHY IS THE HOLE OVER THE ANTARCTIC

Observed ozone over the British Antarctic Survey station at Halley Bay first revealed obvious decreases in the early 1980s compared to data obtained since 1957. The ozone hole is formed each year when there is a sharp decline (currently up to 60 per cent) in the total ozone over most of Antarctica for a period of about two months during southern hemisphere spring (September and October). Man-made emissions of CFCs occur mainly in the northern hemisphere, with about 90 per cent released in Europe, Russia, Japan, and North America. Gases such as CFCs that are insoluble in water and relatively unreactive are mixed throughout the lower atmosphere and rise from the lower atmosphere into the stratosphere; winds then move this air poleward. Normally, chlorine and bromine is inactive, locked up in stable compounds, and does not destroy the ozone.

However, during the Antarctic winter months (June to August) when the region receives no sunlight, the stratosphere becomes cold enough (-80^oC) for high level clouds to form, called Polar Stratospheric Clouds (PSCs). These PSCs provide an ideal catalytic surface on which the chlorine can react with the ozone, thus destroying the ozone layer. This reaction requires sunlight, and therefore only begins when the Sun returns to Antarctica in spring (September to October), before the PSCs have had a chance to melt.

The ozone hole disappears again when the Antarctic air warms up enough during late spring and summer During the southern hemisphere winter, Antarctica is isolated from the rest of the world by a natural circulation of wind called the polar vortex. This prevents atmospheric mixing of stratospheric ozone, thus contributing to the depletion of ozone. Although some ozone depletion occurs over the Arctic, meteorological conditions there are very different to Antarctica and so far have prevented the formation of ozone holes as large as in the southern hemisphere.

THE OZONE HOLE: ARCTIC

Unlike the Arctic, the Antarctic is isolated from the rest of the world during the winter and spring by a natural circulation of wind called the polar vortex. This prevents atmospheric mixing of stratospheric ozone, thus contributing to the depletion of ozone. However, even small depletions in the Arctic region would give cause for considerable concern due to the higher populations in the higher latitudes of the northern hemisphere.

DO OZONE HOLES FORM OVER THE ARCTIC

Arctic ozone depletion has not been as marked as over the Antarctic for two reasons

1. The stratospheric temperatures are seldom below –80°C due to frequent exchange of air masses with the mid latitudes;
2. The Arctic air vortex usually dissipates in late winter before sunlight returns to initiate the ozone destruction.

The differences between the two regions result in part from the larger land mass in the northern hemisphere, which causes more activity in the atmosphere. Nevertheless, analysis of satellite data reveals that the loss of ozone in the northern hemisphere is now proceeding faster than previously thought. In 1989, NASA's Airborne Arctic Stratospheric Expedition, the first comprehensive research expedition to explore the Arctic region, found that the Arctic stratosphere in winter has almost as much chlorine monoxide as is found in Antarctica, the same destructive chlorine that causes the Antarctic ozone hole.

While no Arctic ozone losses comparable with those in the Antarctic have occurred, localised Arctic ozone losses have been observed in winter concurrent with observation of elevated levels of reactive chlorine, made available through man-made emissions of CFCs. Ozone losses have increased greatly in the 1990s in the Arctic and in late 1997 were the greatest ever observed, just as to measurements by NASA satellites. The rate of loss in mid-latitudes has reached 8 per cent per decade in late winter and spring, and significant loss is now encroaching on the growing season. This compares with the yearly-averaged global mean decrease in the amount of ozone of about 3 per cent in the last decade.

CLIMATE CHANGE AND OZONE DEPLETION IN THE ARCTIC

An ozone hole in the Arctic is expected to grow larger over the coming decades as a result of man-made greenhouse gas emissions which may cause climate change, before recovering after 2020. Loss of ozone in the Arctic by 2020 could be about double what would occur without greenhouse gases.

Though greenhouse gases cause atmospheric warming at the Earth's surface, they cool the stratosphere by trapping more heat below, in the troposphere. Since ozone chemistry is very sensitive to temperature, particularly at -80oC when Polar Stratospheric Clouds can form, this stratospheric cooling may result in more ozone depletion in the Arctic.

Temperatures are slightly warmer in the Arctic than the Antarctic during their respective winter and spring seasons, with the result that ozone loss in the northern hemisphere has been lower than that in the southern hemisphere.

But the Arctic stratosphere has gradually cooled over the last decade, resulting in the increased ozone loss. Computer models predict that temperature and wind changes induced by greenhouse gas emissions may allow a stronger and longer-lasting atmospheric vortex to form above the Arctic, as in the Antarctic, causing an increase in ozone depletion. Because of international controls on the emission of ozone-depleting chemicals, those gases are expected to peak about the year 2000.

However, greenhouse gas emissions continue to increase, despite international efforts to control them, and Arctic ozone depletion may continue to worsen until the 2010s, with two-thirds of atmospheric ozone lost in the most severely affected areas.

NATIONAL INFLUENCES ON STRATOSPHERIC OZONE

Despite all of our harmful chemicals going into the atmosphere, some people still argue that stratospheric ozone depletion is just part of a natural variation. Natural variations which influence the amount of ozone in the upper atmosphere include solar activity, volcanic eruptions and changes in atmospheric circulation - the planetary winds.

THE SUN'S INFLUENCE ON OZONE

Stratospheric ozone is primarily created by ultraviolet (UV) radiation coming from the Sun. The Sun's energy release does vary, especially over the 11-year sunspot cycle. During the active phase of the 11-year sunspot cycle, more ozone is produced with the increased UV coming to Earth. This phenomenon can boost the average ozone concentration over the poles by about 4 per cent, but when this is averaged out over the whole earth, the world average ozone increase is about 2 per cent.

Observations since the 1960s have shown that total global ozone levels have decreased by 1-2 per cent from the maximum to the minimum of a typical cycle. However, since downward trends in ozone levels are much larger than 1-2 per cent, particularly at the higher latitudes, the Sun's output cannot be wholly responsible. Unusual solar activity can cause the ozone levels in the upper stratosphere to be substantially depleted, but since most of the ozone is in the middle stratosphere, the effect on the total ozone column is negligible.

ATMOSPHERIC WINDS AND OZONE

A natural cycle in which prevailing tropical winds in the lower stratosphere vary over a time span of about two years can also influence the amount of ozone in the stratosphere.

A change from easterly flow to westerly flow can bring up to a 3 per cent increase in ozone over certain locations, but it is usually cancelled out when the total ozone of the Earth is averaged.

VOLCANIC ERUPTIONS AND OZONE

Volcanic eruptions are one of the few natural things that can have a diminishing effect on the ozone layer. Large eruptions can potentially inject significant quantities of chlorine (via hydrochloric acid - HCl) directly in the stratosphere where the highest concentrations of ozone are found. However, the vast majority of volcanic eruptions are too weak to reach the stratosphere, around 10 km above the surface. Thus, any HCl emitted in the eruption remains in the troposphere where it is quickly dissolved and washed out by rain. In addition, there is no historical record that shows significant increases in chlorine in the stratosphere following even the most major eruptions. It is also possible that ice particles containing sulphuric acid from large volcanic eruptions may contribute to ozone loss. When chlorine compounds resulting from the break-up of man-made CFCs in the stratosphere are present, the sulphate particles serve to convert them into more active forms that may cause more rapid ozone depletion.

In 1991 Mt. Pinatubo in the Philippines erupted tonnes of dust and gas high into the atmosphere which caused global reductions in the ozone layer for 2 to 3 years. Thus, whilst large volcanic eruptions may increase the rate of stratospheric ozone depletion, it is more probable that the presence of chlorine from man-made CFC emissions is the chief cause of ozone loss in the first instance.

MONITORING OZONE DEPLETION

In 1974, after millions of tons of CFCs had been manufactured and sold, chemists F. Sherwood Rowland and Mario Molina of the University of California began to wonder where all these CFCs ended up. Rowland and Molina theorised that short waves of ultraviolet radiation from the Sun in the stratosphere would break up CFCs, and that the free chlorine atoms would then enter into a chain reaction, destroying ozone. Many people, however, remained unconvinced of the danger until the mid-1980s, when a severe annual depletion of ozone was first monitored by the British Antarctic Survey above Antarctica.

The depletion above the South Pole was so severe that the British geophysicist, Joe Farman, who first measured it assumed his spectrophotometer must be broken and sent the device back to England to be repaired. Once the depletion was verified, it came to be known throughout the world through a series of NASA satellite photos as the Antarctic Ozone Hole.

EVIDENCE FOR STRATOSPHERIC OZONE DEPLETION

Laboratory studies, backed by satellite and ground-based measurements, show that chlorine reacts very rapidly with ozone. They also show that the chlorine oxide formed in that reaction undergoes further processes that regenerate the original chlorine, allowing the sequence to be repeated very many times (a "chain reaction"). Similar reactions also take place between bromine and ozone. Many other reactions are often also taking place simultaneously in the stratosphere, making the connections among the changes difficult to untangle. Nevertheless, whenever chlorine (or bromine) and ozone are found together in the stratosphere, the ozone-destroying reactions must be taking place. Observations of the Antarctic ozone hole have given a convincing and unmistakable demonstration of these processes.

MONITORING OF OZONE DEPLETION

There has been much monitoring of the condition of the ozone layer in the last decade since the Antarctic ozone hole was first discovered by the British Antarctic Survey. This has utilised satellites and other groundbased resources that are dedicated to observing the destruction of stratospheric ozone. The main satellite that monitors the ozone is the TOMS (Total Ozone Mapping Spectrometer) satellite. The TOMS satellite measures the ozone levels from the backscattered sunlight in the ultraviolet (UV) range. Another satellite is NASA's UARS (Upper Atmosphere Research Satellite) which was launched in September 1991. This satellite is unique because it was configured to not only measure ozone levels, but also levels of ozone-depleting chemicals. GOME, launched in April 1995 on the ERS-2 satellite, marks the beginning of a long-term European ozone monitoring effort. Scientists expect to receive high quality data on the global distribution of ozone and several other climateinfluencing trace gases in the Earth's atmosphere.

The German Neumayer Antarctic Research Station was completed in March of 1992, which is located on the Ekstsoem Ice Shelf. This ground-based station studies geographical, meteorological, and air chemistry conditions. In 1987, Canada became the first country in the world to focus on the Arctic ozone layer, following the discovery of the ozone hole over the Antarctic. A cross-country network of monitoring stations has kept continuous watch on Canada's ozone layer for more than three decades. The existence of these early records, before any major human influence on the upper atmosphere, is vital to understanding the changes that have occurred in the ozone layer. In the UK, stratospheric ozone levels are monitored every winter and spring at Cambourne, in Cornwall and Lerwick, in the Shetland Isles.

CURRENT UNDERSTANDING OF PAST OZONE CHANGES

A number of processes were described that can influence the distribution of and changes in stratospheric ozone. Here we discuss the current understanding of past ozone changes, based in part on numerical models that attempt to include these processes. A hierarchy of models of increasing sophistication is used. Some models include relatively few processes whereas others attempt to include many more. The models have their different uses, and strengths. The material here relies heavily on that report, which deals with these topics in much greater detail.

MID-LATITUDE OZONE DEPLETION

In the upper stratosphere the observed ozone depletion over the last 25 years is statistically robust in the sense that its value is not sensitive to small differences in the choice of the time period analysed. This behaviour accords with the fact that in this region of the atmosphere, ozone is under photochemical control and thus only weakly affected by dynamical variability. We therefore expect upper-stratospheric ozone abundance to reflect long-term changes in temperature and in the abundance of species that react chemically with ozone. Over the past 25 years there have been large changes in the abundance of halogen compounds, and the extent of the observed ozone decrease is consistent with the observed increase in anthropogenic chlorine, as originally predicted by Molina and Rowland and Crutzen.

In particular, the vertical and latitudinal profiles of ozone trends in the upper stratosphere are reproduced by 2-D photochemistry models. In the upper stratosphere the attribution of ozone loss in the last couple of decades to anthropogenic halogens is clear-cut. The 2-D models indicate that changes in halogens make the largest contribution to the observed loss of about 7 per cent per decade. The observed cooling of the upper stratosphere will have reduced the rate of ozone destruction in this region. If we take the observed variation of ozone with temperature in the upper stratosphere, then a cooling of 2 K per decade should have led to an increase in ozone of about 2 per cent per decade, partially compensating the loss caused by halogens. Note, however, that in the upper stratosphere radiative- transfer models suggest roughly equal contributions to the cooling from ozone decreases and carbon dioxide increases; the changes in the ozone-temperature system are non-linear. Changes in CH_4 and N_2O will have also played a small role here. Note also that although 2-D models have been widely used in many ozone assessments, they have important limitations in terms of, for example, their ability to include the full range of feedbacks between chemistry and dynamics, or their ability to reproduce the polar vortex.

They are expected to be most accurate in the upper stratosphere, where dynamical effects on ozone are weakest. However, the changes in upper-stratospheric ozone represent only a small contribution to the total changes in column ozone observed over the last 25 years, except in the tropics where there is no statistically significant trend in column ozone. Most of the column ozone depletion in the extratropics occurs in the lower stratosphere, where the photochemical time scale for ozone becomes long and the ozone distribution is sensitive to dynamical variability as well as to chemical processes. This complicates the problem of attribution of the observed ozone decline. Strong ozone variability and complex coupled interactions between

Ozone Depletion

dynamics and chemistry, which are not separable in a simple manner, make attribution particularly difficult.

A number of 2-D models contributed of WMO and a schematic showing their simulations of column ozone between 60°S and 60°N from 1980 to 2050. As there are no significant trends in tropical ozone, the observed and modelled changes are attributable to mid-latitudes. The shaded area shows the results of 2-D photochemistry models forced by observed changes in halocarbons, other source gases and aerosols from 1980 to 2000. Overall, the models broadly reproduce the long-term changes in mid-latitude column ozone for 1980–2000, within the range of uncertainties of the observations and the model range. The spread in the model results comes mainly from their large spread over the SH mid-latitudes, and is at least partly a result of their treatment of the Antarctic ozone hole. In addition, the agreement between models and observations over 60°S to 60°N hides some important disagreements within each hemisphere. In particular, models suggest that the chemical signal of ozone loss following the major eruption of the Mt. Pinatubo volcano in the early 1990s should have been symmetric between the hemispheres, but observations show a large degree of inter-hemispheric asymmetry in mid- latitudes. Changes in atmospheric dynamics can also have a significant influence on NH mid-latitude column ozone on decadal time scales. Natural are all likely to contribute to these dynamical changes. Furthermore, because chemical and dynamical processes are coupled, their contributions to ozone changes cannot be considered in isolation. This coupling is especially complex and difficult to understand with regard to dynamical changes in the tropopause region.

There is an observed relationship between column ozone and several tropospheric circulation indices, including tropopause height. Over time scales of up to about one month, it is the dynamical changes that cause the ozone changes, whereas on longer time scales feedbacks occur and the causality in the relationship becomes unclear. Thus, although various tropospheric circulation indices have changed over the last 20 years in the NH in such a way as to imply a decrease in column ozone, this inference is based on an extrapolation of short-time-scale correlations to longer time scales, which may not be valid. The tropopause, stratospheric PWD drives the seasonal winter- spring ozone build-up in the extratropics, and has essentially no interannual memory. It follows that the observed decrease in NH PWD in the late winter and spring has likely contributed to the observed decrease in NH column ozone over the last 20 to 25 years.

The effect of changes in PWD on the ozone distribution is understood in general terms, but its quantification in observations is relatively crude.

The seasonality of the long-term changes in mid-latitude column ozone differs between hemispheres. In the NH, the maximum decrease is found in spring, and it decays through to late autumn. Fioletov and Shepherd have shown that in the NH the ozone decreases in summer and early autumn are the photochemically damped signal of winter-spring losses, and thus arise from the winter-spring losses without any need for perturbed chemistry in the summertime. The same seasonality is not enough for explaining the summer and early autumn ozone decreases observed in the SH mid-latitudes because they are comparable with the winter-spring losses and therefore point to the influence of transport of ozone-depleted air into mid-latitudes following the break-up of the ozone hole.

In summary, the vertical, latitudinal and seasonal characteristics of past changes in mid-latitude ozone are broadly consistent with the understanding that halogens are the primary cause of these changes. However, to account for decadal variations it is necessary to include consideration of the interplay between dynamical and chemical effects as well as the impact of variations in aerosol loading; our inadequate quantitative understanding of these processes limits our predictive capability.

WINTER-SPRING POLAR DEPLETION

Polar ozone depletion in the winter-spring period is generally considered separately from mid-latitude depletion, because of the extremely severe depletion that can occur in polar regions from heterogeneous chemistry on PSCs. We consider first the Antarctic, and then the Arctic. The Antarctic ozone hole represents the most striking example of ozone depletion in the atmosphere. It developed through the 1980s as chlorine loading increased, and has recurred every year since then. The Antarctic ozone hole has been clearly attributed to anthropogenic chlorine through field campaigns and photochemical modelling.s

An ozone hole now occurs every year because, in addition to the availability of anthropogenic chlorine, wintertime temperatures in the Antarctic lower stratosphere are always low enough for chlorine activation to occur on PSCs prior to the return of sunlight to the vortex in the spring, and the air is always sufficiently isolated within the vortex for it to become strongly depleted of ozone. There is nevertheless dynamically induced variability in the extent and severity of the ozone hole. The most dramatic instance of such variability occurred in 2002, when the Antarctic stratosphere experienced its first observed sudden warming. The sudden warming split the vortex in two and halted the development of the ozone hole that year. Such events occur commonly in the NH, as a result of the stronger planetary-wave forcing in the NH, but had never before been seen in the SH, although there had been previous instances of disturbed winters.

Although unprecedented, this event resulted from dynamical variability and was not indicative of ozone recovery; indeed, the ozone hole in 2003 was back to a severity characteristic of the 1990s. Given the fairly predictable nature of the Antarctic ozone hole, its simulation constitutes a basic test for models. Twodimensional models are not expected to represent the ozone hole well because they cannot properly represent isolation of polar vortex air and their estimates of Antarctic ozone loss vary greatly. However, an emerging tool for attribution is the 3-D chemistry-climate model. These models include an on- line feedback of chemical composition to the dynamical and radiative components of the model. The CCMs currently available have been developed with an emphasis on the troposphere and stratosphere. They consider a wide range of chemical, dynamical and radiative feedback processes and can be used to address the question of why changes in stratospheric dynamics and chemistry may have occurred as a result of anthropogenic forcing. The models specify WMGHGs, source gases, aerosols and often SSTs, but otherwise run freely and exhibit considerable interannual variability. Thus in addition to issues of model accuracy, it is necessary to consider how representative are the model simulations. In recent years a number of CCMs have been employed to examine the effects of climate change on ozone. They have been used for long-term simulations to try to reproduce observed past changes, such as ozone and temperature trends. The models have been run either with fixed forcings to investigate the subsequent 'equilibrium climate' or with time-varying forcings.

atmospheric features with respect to the mean conditions and the seasonal and interannual variability; the long-term changes in the dynamical and chemical composition of the upper troposphere and the stratosphere are also in reasonable agreement with observations. But a reasonable reproduction of the timing of the ozone hole and an adequate description of total column ozone do not necessarily mean that all processes in the CCMs are correctly captured. There are obvious discrepancies between the models themselves, and between model results and detailed observations. Some of the differences among the models are caused by the fact that the specific model systems employed differ considerably, not only in complexity but also in the vertical extent of the model domain and in the horizontal and vertical resolutions. An important difference between models and observations originates to a large extent from the cold bias that is found in the high latitudes of many CCMs, particularly near the tropopause and in the lower stratosphere.

The low-temperature bias of the models is generally largest in winter over the South Pole, where it is of the order of 5–10 K; this magnitude of bias could be significant in controlling planetary- wave dynamics and restricting the interannual variability of the models. A low-temperature bias in the lower stratosphere of a model has a significant impact not only on model heterogeneous chemistry but also on the transport of chemical species and its potential change due to changes in circulation. Moreover, changes in stratospheric dynamics alter the conditions for wave forcing and wave propagation which in turn influence the seasonal and interannual variability of the atmosphere. The reasons for the cold bias in the models are still unknown. This bias has been reduced in some models, for example by considering non- orographic gravity-wave drag schemes but the problem is not yet solved. Without its resolution, the reliability of these models for attribution and prediction is reduced. Despite these potential problems, CCMs simulate the development of the Antarctic ozone hole reasonably well. The models examined in the intercomparison of Austin et al. agree with observations of minimum Antarctic springtime column ozone over 1980–2000 within the model variability, confirming that the Antarctic ozone hole is indeed a robust response to anthropogenic chlorine. The cold-pole biases of some models seem not to affect minimum column ozone too much, although some modelled minima are significantly lower than observed.

None of the models exhibited a sudden warming as seen in 2002, although several exhibit years with a disturbed vortex, as seen in their interannual variability in minimum column ozone. However, minimum total ozone is not the only diagnostic of the ozone hole. Many models significantly underestimate the area of the ozone hole. In contrast with the Antarctic, winter-spring ozone abundance in the Arctic exhibits significant year-to-year variability. This variability arises from the highly disturbed nature of Arctic stratospheric dynamics, including relatively frequent sudden warmings. Because dynamical variability affects both transport and temperature, dynamical and chemical effects on ozone are coupled.

The dynamical variability is controlled by stratospheric PWD; years with strong PWD, compared with years with weak PWD, have stronger downwelling over the pole, and thus higher temperatures, a weaker vortex and more ozone transport. Because low temperatures and a stronger vortex tend to favour chemical ozone loss in the presence of elevated halogen levels, it follows that dynamical and chemical effects tend to act in concert. In the warmest years there is essentially no chemical ozone loss and ozone levels are similar to those seen pre-1980. The chemical ozone loss was calculated for cold years using various methods of WMO, and the agreement between the methods provided confidence in the estimates.

The calculations showed a roughly linear relationship between chemical ozone loss and temperature. Furthermore, the in situ chemical ozone loss was found to account for roughly one half of the observed ozone decrease between cold years and warm years. Given the high confidence in the estimates of chemical ozone loss, the remaining half can be attributed to reduced ozone transport, as is expected in cold years with weak PWD, a strong vortex, and weak downwelling over the pole. From this it can be concluded that in cold years, chlorine chemistry doubled the ozone decrease that would have occurred from transport alone. Compared with other recent decades, the 1990s had an unusually high number of cold years, and these led to low values of Arctic ozone. In more recent years, Arctic ozone has been generally higher, although still apparently below pre- 1980 values. This behaviour does not reflect the time evolution of stratospheric chlorine loading. Rather, it reflects the meteorological variability in the presence of chlorine loading. In this respect the Arctic ozone record needs to be interpreted in the context of the meteorology of a given year, far more than is generally the case in the Antarctic.

This interpretation of Arctic ozone changes is consistent with the fact that the observed decrease in Arctic stratospheric temperature over 1980–2000 cannot be explained from direct radiative forcing due to ozone depletion or changes in green- house gases alone, although they do make a contribution. The inference is that the observed springtime cooling was mainly the result of decadal meteorological variability in PWD. This is in contrast with the Antarctic, where the observed cooling in November and the prolonged persistence of the vortex have been shown to be the result of the ozone hole. There is seen to be little impact from CO_2 at these altitudes over this time period.

In the Arctic, the cooling induced by ozone loss is too small and too late in the season to account for the observed cooling. Rather, the observed cooling is required in order to initiate severe Arctic ozone depletion. It is not known what has caused the recent decadal variations in Arctic temperature. Rex et al. have argued that the value of VPSC in cold years has systematically increased since the 1960s.

However, the Arctic exhibits significant decadal variability and it is not possible to exclude natural variability as the cause of these changes. These considerations have implications for the attribution of past changes. No matter how good a CCM is and how well its climate-change experiments are characterized, it cannot exactly reproduce the real atmosphere because the real atmosphere is only one possible realization of a chaotic system. The best one can expect, even for a perfect model is that the observations fall within the range of model-predicted behaviours, just as to appropriate statistical criteria. Whether this permits a meaningful prediction depends on the relative magnitudes and time scales of the forced signal and the natural noise. Whereas the evolution of Antarctic ozone is expected to be fairly predictable over decadal time scales, it is not at all clear whether this is the case in the Arctic.

Thus, it is not a priori obvious that even a perfect CCM would reproduce the decreases in Arctic ozone observed over the past 20 years, for example. Bearing these caveats in mind, the simulations of past Arctic minimum ozone from the CCMs considered in Austin et al. The models seem generally to have a positive bias with respect to the observations over the same period. None of the CCMs achieve column ozone values as low as those observed, and the modelled ozone trends are generally smaller than the observed trend. However, the range of variability exhibited by each of the models is considerable, and it is therefore difficult to say that the models are definitely deficient on the basis of the ozone behaviour alone; more detailed diagnostics are required and with observations of other chemical species. In summary, the development of the Antarctic ozone hole through the 1980s was a direct response to increasing chlorine loading, and the severity of the ozone decrease has not changed since the early 1990s, although there are year-to-year variations.

In the Arctic, the extent of chemical ozone loss due to chlorine depends on the meteorology of a given winter. In winters that are cold enough for the existence of PSCs, chemical ozone loss has been identified and acts in concert with reduced transport to give Arctic ozone depletion. There was particularly severe Arctic ozone depletion in many years of the 1990s as a result of a series of particularly cold winters. It is not known whether this period of low Arctic winter temperatures is just natural variability or a response to changes in greenhouse gases.

THE MONTREAL PROTOCOL, FUTURE OZONE CHANGES AND THEIR LINKS TO CLIMATE

The halogen loading of the stratosphere increased rapidly in the 1970s and 1980s. As a result of the Montreal Protocol and its Amendments and Adjustments, the stratospheric loading of chlorine and bromine is expected to decrease slowly in the coming decades, reaching pre-1980 levels some time around 2050. If chlorine and bromine were the only factors affecting stratosphere ozone, we would then expect stratospheric ozone to 'recover' at about the same time. Over this long time scale, the state of the stratosphere may well change because of other anthropogenic effects, in ways that affect ozone abundance. For example, increasing concentrations of CO_2 are expected to further cool the stratosphere, and therefore to influence the rates of ozone destruction.

Ozone Depletion 81

Any changes in stratospheric water vapour, CH4 and N2O, all of which are difficult to predict quantitatively, will also affect stratospheric chemistry and radiation. In addition, natural climate variability including, volcanic eruptions, can affect ozone on decadal time scales. For these reasons, 'recovery' of stratospheric ozone is a complicated issue. A number of model calculations to investigate recovery were reported in WMO, using both 2-D and 3-D models. The main results from that assessment, and new studies reported since then, are summarized here. As with the discussion of past ozone changes, we first discuss mid-latitude changes and then polar changes.

Mid-Latitude Ozone

Predictions of future mid-latitude ozone change in response to expected decreasing halogen levels have been extensively discussed in WMO based primarily on simulations with eight separate 2-D photochemistry models that incorporated predicted future changes in halogen loading. Several scenarios for future changes in trace climate gases were also incorporated in these simulations, although the ozone results were not particularly sensitive to which scenario was employed. The models generally predict a minimum in global column ozone in 1992–1993 following the eruption of Mt. Pinatubo, followed by steady increases. The latter evolution is primarily determined by changes in atmospheric chlorine and bromine loading, which reaches a maximum in approximately 1995 and then slowly decreases.

There is, however, a large spread in the times the models predict that global ozone will return to 1980 levels, ranging from 2025 to after 2050. The spread of results arises in part from the differences between the models used. Two of the 2-D models used in the simulations included interactive temperature changes caused by increasing greenhouse gases, which result in a long-term cooling of the stratosphere. Cooling in the upper stratosphere results in slowing of the gas-phase chemical cycles that destroy ozone there and consequently these two models show a greater increase with time of total ozone than the other models. An important caveat is that most of the past column ozone change has occurred in the lower stratosphere, and future temperature feedback effects are more complicated and uncertain in the lower stratosphere. Furthermore, these effects involve changes in transport, heterogeneous chemistry and polar processes, which are not accurately simulated in 2-D models. At present there is considerable uncertainty regarding the details of temperature feedbacks on future midlatitude column ozone changes. Because 2-D models can be expected to capture the main chemical processes involved in ozone recovery, reliable predictions of future changes in the coupled ozone-climate system require the use of CCMs, since they consider possible changes in climate feedback mechanisms. Despite some present limitations, these models have been employed in sensitivity studies to assess the global future development of the chemical composition of the stratosphere and climate.

The results have been compared with each other to assess the uncertainties of such predictions and have been documented in international assessment reports. However most of the attention has been focused on polar ozone. Two dynamical influences appear to be related to NH mid-latitude ozone decreases over 1980–2000: a decrease in PWD and an increase in tropopause height.

These mechanisms likewise have the potential to influence future ozone. If the past dynamical changes represent natural variability, then one cannot extrapolate past dynamical trends, and ozone recovery could be either hastened or delayed by dynamical variability. If, on the other hand, the past dynamical changes represent the dynamical response to WMGHG-induced climate change, then one might expect these changes to increase in magnitude in the future. This would decrease future ozone levels and delay ozone recovery.

In the case of tropopause height changes, to the extent that they are themselves caused by ozone depletion, the changes should reverse in the future. Another potential dynamical influence on ozone arises via water vapour. Stratospheric water vapour is controlled by two processes: methane oxidation within the stratosphere and the transport of water vapour into the stratosphere. The latter depends in large part on the temperature of the tropical tropopause, although the precise details of this relationship remain unclear. In the future, changes in tropical tropopause temperature could conceivably affect the water vapour content of the stratosphere. Unfortunately, our ability to predict future dynamical influences on ozone is very poor, for two reasons. First, dynamical processes affecting ozone exhibit significant temporal variability and are highly sensitive to other aspects of the atmospheric circulation.

This means that the WMGHG-induced signal is inherently difficult to isolate or to represent accurately in CCMs, whereas the climate noise is relatively high. Second, there are still uncertainties in the performance of the CCMs, which are the tools needed to address this question. Predictions of WMGHG-induced dynamical changes by CCMs do not even agree on the sign of the changes: some models predict a weakened PWD, which would decrease mid-latitude ozone by weakening transport, whereas others predict a strengthened PWD. Note that these dynamical changes would also affect the lifetime of stratospheric pollutants, with stronger PWD leading to shorter lifetimes Thus, the direct effect of PWD change on ozone transport would be reinforced over decadal time scales by altered chemical ozone loss arising from PWD-induced changes in stratospheric chlorine loading. To summarize

Ozone Depletion

the results from 2-D models, upper-stratospheric ozone in the mid-latitudes will recover to pre-1980 levels well before stratospheric chlorine loading returns to pre- 1980 levels, and could even overshoot pre-1980 levels because of CO_2- induced upper-stratospheric cooling. However, details of this future evolution are sensitive to many factors and there is significant divergence between different models. In terms of mid-latitude column ozone, where changes are expected to be dominated by ozone in the lower stratosphere, 2-D models all predict a steadily increasing ozone abundance as halogen levels decrease.

While these models usually include a detailed treatment of polar chemistry, they do not include a detailed treatment of polar dynamics and mixing to mid-latitudes, or of dynamical aspects of climate change. Any future changes in stratospheric circulation and transport, or a large Pinatubo-like volcanic eruption, could have the potential to affect global column ozone, both directly and indirectly via chemical processes. Quantitative prediction is clearly difficult. At this stage, CCM simulations provide sensitivity calculations, which allow for the exploration of some examples of possible future evolution.

Polar Ozone

As noted earlier, 2-D models do not provide a realistic treatment of polar processes, so predictions of polar ozone rely principally on CCMs. An intensive inter-comparison of CCM predictions, the first of its kind, was performed of WMO. It is important to recognize that, apart from model deficiencies, CCM predictions are themselves subject to model variability and thus must be viewed in a statistical sense. Because of computer limitations, a large number of simulations with a single CCM cannot be carried out, so the models cannot yet be generally employed for ensemble runs. The approach taken in Austin et al. was to regard the collection of different CCMs as representing an ensemble. The collection included, moreover, a mixture of transient and time- slice runs.

For the Antarctic, where the CCMs all reproduce the development of the ozone hole in a reasonably realistic manner. Considering all the models, there is an overall consensus that the recovery of the Antarctic ozone hole will essentially follow the stratospheric chlorine loading. There is a hint in of a slight delay in the recovery compared with the peak in chlorine loading, presumably because of a cooling arising from the specified increase in WMGHG concentrations. Austin et al. estimated that the recovery of the Antarctic ozone layer can be expected to begin any year within the range 2001 to 2008. This would mean that Antarctic ozone depletion could slightly increase within the next few years despite a decrease in stratospheric chlorine loading, because lower temperatures would increase chlorine activation. However there is considerable natural variability, so it is also quite possible that the most severe Antarctic ozone hole has already occurred. In the Arctic, where ozone depletion is more sensitive to meteorological conditions, the picture drawn by the CCMs is not consistent. Whereas most CCMs also predict a delayed start of Arctic ozone recovery, others indicate a different development: they simulate an enhanced PWD that produces a more disturbed and warmer NH stratospheric vortex in the future. This 'dynamical heating' more than compensates for the radiative cooling due to enhanced greenhouse-gas concentrations.

Under these circumstances, and in combination with reduced stratospheric chlorine concentrations, polar ozone in these CCMs recovers within the next decade to values measured before the start of stratospheric ozone depletion. However, none of the current models suggest that an 'ozone hole', similar to that observed in the Antarctic, will occur over the Arctic. Analyses of model results show that the choice of the prescribed SSTs seems to play a critical role in the planetary-wave forcing. Predicted SSTs vary in response to the same forcings that are already included in CCMs. Realistic predictions of future SSTs are therefore a necessary prerequisite for stratospheric circulation predictions, at least in the NH extratropics. The differences between the currently available CCM results clearly indicate the uncertainties in the assessment of the future development of polar ozone. In the NH the differences are more pronounced than in the SH. Part of these differences arises from the highly variable nature of the NH circulation, as is reflected in the past record.

Until now, the different transient model predictions have been based on single realizations, not ensembles, so it is not yet possible to determine whether the differences between the models. The time-slice simulations give some indication of the range of natural variability that is possible, and allow for a wide range of future possibilities. Nevertheless there are also significant uncertainties arising from the performance of the underlying dynamical models. Cold biases have been found in the stratosphere of many of the models, with obvious direct consequences for chemistry. In consequence, at the current stage of model development, uncertainties in the details of PSC formation and sedimentation might be less important than model temperature biases for simulating accurate ozone amounts, although it is possible that the models underestimate the sensitivity of chemical ozone loss to temperature changes. However, because it has been shown that denitrification does contribute significantly to Arctic ozone loss and is very sensitive to temperature, both problems have to be solved before a reliable estimate of future Arctic ozone losses is possible. Further investigations and model developments, combined with data analysis, are needed in order to reduce or eliminate these various model deficiencies and quantify the natural variability. In summary, the Antarctic ozone hole is expected to recover more or less following the decrease in chlorine loading, and return to 1980 levels in the 2045 to 2055 time frame. There may be a slight delay arising from WMGHG-induced cooling, but natural variability in the extent of the ozone hole is sufficiently large that the most severe ozone hole may have already occurred, or may occur in the next five years or so.

With regard to the Arctic, the future evolution of ozone is potentially sensitive to climate change and to natural variability, and will not necessarily follow strictly the chlorine loading. There is uncertainty in even the sign of the dynamical feedback to WMGHG changes. Numerical models like CCMs are needed to make sensitivity studies to estimate possible future changes, although they are currently not fully evaluated and their deficiencies are obvious. Therefore, the interpretation of such 'predictions' must be performed with care. Progress will result from further development of CCMs and from comparisons of results between models and with observations. This will help to get a better understanding of potential feedbacks in the atmosphere, thereby leading to more reliable estimates of future changes.

Chapter 3

Climate Change

Climate change is a long-term change in the statistical distribution of weather patterns over periods of time that range from decades to millions of years. It may be a change in the average weather conditions or a change in the distribution of weather events with respect to an average, for example, greater or fewer extreme weather events. Climate change may be limited to a specific region, or may occur across the whole Earth. In recent usage, especially in the context of environmental policy, climate change usually refers to changes in modern climate. It may be qualified as anthropogenic climate change, more generally known as global warming or anthropogenic global warming (AGW).

TERMINOLOGY

The most general definition of climate change is a change in the statistical properties of the climate system when considered over periods of decades or longer, regardless of cause. Fluctuations on periods shorter than a few decades, such as El Nino, do not represent climate change. The term sometimes is used to refer specifically to climate change caused by human activity; for example, the United Nations Framework Convention on Climate Change defines climate change as "a change of climate which is attributed directly or indirectly to human activity that alters the composition of the global atmosphere and which is in addition to natural climate variability observed over comparable time periods." In the latter sense climate change is synonymous with global warming.

CAUSES

Factors that can shape climate are climate forcings. These include such processes as variations in solar radiation, deviations in the Earth's orbit, mountain-building and continental drift, and changes in greenhouse gas concentrations. There are a variety of climate change feedbacks that can either amplify or diminish the initial forcing. Some parts of the climate system, such as the oceans and ice caps, respond slowly in reaction to climate forcing because of their large mass. Therefore, the climate system can take centuries or longer to fully respond to new external forcings.

PLATE TECTONICS

Over the course of millions of years, the motion of tectonic plates reconfigures global land and ocean areas and generates topography. This can affect both global and local patterns of climate and atmosphere-ocean circulation. The position of the continents determines the geometry of the oceans and therefore influences patterns of ocean circulation. The locations of the seas are important in controlling the transfer of heat and moisture across the globe, and therefore, in determining global climate. A recent example of tectonic control on ocean circulation is the formation of the Isthmus of Panama about 5 million years ago, which shut off direct mixing between the Atlantic and Pacific Oceans.

This strongly affected the ocean dynamics of what is now the Gulf Stream and may have led to Northern Hemisphere ice cover. During the Carboniferous period, about 300 to 360 million years ago, plate tectonics may have triggered large-scale storage of carbon and increased glaciation. Geologic evidence points to a "megamonsoonal" circulation pattern during the time of the supercontinent Pangaea, and climate modeling suggests that the existence of the supercontinent was conducive to the establishment of monsoons. The size of continents is also important. Because of the stabilizing effect of the oceans on temperature, yearly temperature variations are generally lower in coastal areas than they are inland. A larger supercontinent will therefore have more area in which climate is strongly seasonal than will several smaller continents or islands.

SOLAR OUTPUT

The sun is the predominant source for energy input to the Earth. Both long- and short-term variations in solar intensity are known to affect global climate. Three to four billion years ago the sun emitted only 70 per cent as much power as it does today. If the atmospheric composition had been the same as today, liquid water should not have existed on Earth. However, there is evidence for the presence of water on the early Earth, in the Hadean and Archean eons, leading to what is known as the faint young sun paradox. Hypothesized solutions to this paradox include a vastly different atmosphere, with much higher concentrations of greenhouse gases than currently exist Over the following approximately 4 billion years, the energy output of the sun increased and atmospheric composition changed, with the oxygenation of the atmosphere around 2.4 billion years ago being the most notable alteration. These changes in luminosity, and the sun's ultimate death as it becomes a red giant and then a white dwarf, will have large effects on climate, with the red giant phase possibly ending life on Earth. Solar output also varies on shorter time scales, including the 11-year solar cycle and longer-term modulations. Solar intensity variations are considered to have been influential in triggering the Little Ice Age, and some of the warming observed from 1900 to 1950. The cyclical nature of the sun's energy output is not yet fully understood; it differs from the very slow change that is happening within the sun as it ages and evolves. While most research indicates solar variability has induced a small cooling effect from 1750 to the present, a few studies point towards solar radiation increases from cyclical sunspot activity affecting global warming.

ORBITAL VARIATIONS

Slight variations in Earth's orbit lead to changes in the seasonal distribution of sunlight reaching the Earth's surface and how it is distributed across the globe. There is very little change to the area-averaged annually averaged sunshine; but there can be strong changes in the geographical and seasonal distribution.

The three types of orbital variations are variations in Earth's eccentricity, changes in the tilt angle of Earth's axis of rotation, and precession of Earth's axis. Combined together, these produce Milankovitch cycles which have a large impact on climate and are notable for their correlation to glacial and interglacial periods, their correlation with the advance and retreat of the Sahara, and for their appearance in the stratigraphic record.

VOLCANISM

Volcanism is a process of conveying material from the crust and mantle of the Earth to its surface. Volcanic eruptions, geysers, and hot springs, are examples of volcanic processes which release gases and/or particulates into the atmosphere. Eruptions large enough to affect climate occur on average several times per century, and cause cooling (by partially blocking the transmission of solar radiation to the Earth's surface) for a period of a few years. The eruption of Mount Pinatubo in 1991, the second largest terrestrial eruption of the 20th century (after the 1912 eruption of Novarupta) affected the climate substantially. Global temperatures decreased by about 0.5 °C (0.9 °F). The eruption of Mount Tambora in 1815 caused the Year Without a Summer. Much larger eruptions, known as large igneous provinces, occur only a few times every hundred million years, but may cause global warming and mass extinctions. Volcanoes are also part of the extended carbon cycle. Over very long (geological) time periods, they release carbon dioxide from the Earth's crust and mantle, counteracting the uptake by sedimentary rocks and other geological carbon dioxide sinks. The US Geological Survey, however, estimates are that human activities generate more than 130 times the amount of carbon dioxide emitted by volcanoes.

OCEAN VARIABILITY

The ocean is a fundamental part of the climate system. Short-term fluctuations (years to a few decades) such as the El Nino–Southern Oscillation, the Pacific decadal oscillation, the North Atlantic oscillation, and the Arctic oscillation, represent climate variability rather than climate change. On longer time scales, alterations to ocean processes such as thermohaline circulation play a key role in redistributing heat by carrying out a very slow and extremely deep movement of water, and the long-term redistribution of heat in the world's oceans.

HUMAN INFLUENCES

In the context of climate variation, anthropogenic factors are human activities that change the environment. In some cases the chain of causality of human influence on the climate is direct and unambiguous (for example, the effects of irrigation on local humidity), while in other instances it is less clear. Various hypotheses for human-induced climate change have been argued for many years. Presently the scientific consensus on climate change is that human activity is very likely the cause for the rapid increase in global average temperatures over the past several decades. Consequently, the debate has largely shifted onto ways to reduce further human impact and to find ways to adapt to change that has already occurred.

Of most concern in these anthropogenic factors is the increase in CO_2 levels due to emissions from fossil fuel combustion, followed by aerosols (particulate matter in the atmosphere) and cement manufacture. Other factors, including land use, ozone depletion, animal agriculture and deforestation, are also of concern in the roles they play - both separately and in conjunction with other factors - in affecting climate, microclimate, and measures of climate variables.

ASTRONOMICAL THEORY AND ABRUPT CLIMATE CHANGES

Throughout the 19th and 20th centuries, a wide range of geomorphology and palaeontology studies has provided new insight into the Earth's past climates, covering periods of hundreds of millions of years. The Palaeozoic Era, beginning 600 Ma, displayed evidence of both warmer and colder climatic conditions than the present; the Tertiary Period was generally warmer; and the Quaternary Period showed oscillations between glacial and interglacial conditions. Louis Agassiz developed the hypothesis that Europe had experienced past glacial ages, and there has since been a growing awareness that long-term climate observations can advance the understanding of the physical mechanisms affecting climate change. The scientific study of one such mechanism–modifications in the geographical and temporal patterns of solar energy reaching the Earth's surface due to changes in the Earth's orbital parametres–has a long history. The pioneering contributions of Milankovitch to this astronomical theory of climate change are widely known, and the historical review of Imbrie and Imbrie calls attention to much earlier contributions, such as those of James Croll, originating in 1864.

The pace of palaeoclimatic research has accelerated over recent decades. Quantitative and well-dated records of climate fluctuations over the last 100 kyr have brought a more comprehensive view of how climate changes occur, as well as the means to test elements of the astronomical theory. By the 1950s, studies of deep-sea cores suggested that the ocean temperatures may have been different during glacial times. Ewing and Donn proposed that changes in ocean circulation actually could initiate an ice age. In the 1960s, the works of Emiliani and Shackleton showed the potential of isotopic measurements in deepsea sediments to help explain Quaternary changes. In the 1970s, it became possible to analyse a deep-sea core time series of more than 700 kyr, thereby using the last reversal of the Earth's magnetic field to establish a dated chronology. This deep-sea observational record clearly showed the same periodicities found in the astronomical forcing, immediately providing strong support to Milankovitch's theory.

Ice cores provide key information about past climates, including surface temperatures and atmospheric chemical composition. The bubbles sealed in the ice are the only available samples of these past atmospheres. The first deep ice cores from Vostok in Antarctica provided additional evidence of the role of astronomical forcing. They also revealed a highly correlated evolution of temperature changes and atmospheric composition, which was subsequently confirmed over the past 400 kyr and now extends to almost

1 Myr. This discovery drove research to understand the causal links between greenhouse gases and climate change.

The same data that confirmed the astronomical theory also revealed its limits: a linear response of the climate system to astronomical forcing could not explain entirely the observed fluctuations of rapid ice-age terminations preceded by longer cycles of glaciations. The importance of other sources of climate variability was heightened by the discovery of abrupt climate changes. In this context, 'abrupt' designates regional events of large amplitude, typically a few degrees celsius, which occurred within several decades–much shorter than the thousand-year time scales that characterise changes in astronomical forcing. Abrupt temperature changes were first revealed by the analysis of deep ice cores from Greenland.

Oeschger et al. recognised that the abrupt changes during the termination of the last ice age correlated with cooling in Gerzensee and suggested that regime shifts in the Atlantic Ocean circulation were causing these widespread changes. The synthesis of palaeoclimatic observations by Broecker and Denton invigourated the community over the next decade. By the end of the 1990s, it became clear that the abrupt climate changes during the last ice age, particularly in the North Atlantic regions as found in the Greenland ice cores, were numerous, indeed abrupt and of large amplitude. They are now referred to as Dansgaard-Oeschger events. A similar variability is seen in the North Atlantic Ocean, with north-south oscillations of the polar front and associated changes in ocean temperature and salinity. With no obvious external forcing, these changes are thought to be manifestations of the internal variability of the climate system.

The importance of internal variability and processes was reinforced in the early 1990s with analysis of records with high temporal resolution. New ice cores new ocean cores from regions with high sedimentation rates, as well as lacustrine sediments and cave stalagmites produced additional evidence for unforced climate changes, and revealed a large number of abrupt changes in many regions throughout the last glacial cycle. Long sediment cores from the deep ocean were used to reconstruct the thermohaline circulation connecting deep and surface waters and to demonstrate the participation of the ocean in these abrupt climate changes during glacial periods.

By the end of the 1990s, palaeoclimate proxies for a range of climate observations had expanded greatly. The analysis of deep corals provided indicators for nutrient content and mass exchange from the surface to deep water, showing abrupt variations characterised by synchronous changes in surface and deep-water properties. Precise measurements of the CH_4 abundances in polar ice cores showed that they changed in concert with the Dansgaard-Oeschger events and thus allowed for synchronisation of the dating across ice cores. The characteristics of the antarctic temperature variations and their relation to the Dansgaard-Oeschger events in Greenland were consistent with the simple concept of a bipolar seesaw caused by changes in the thermohaline circulation of the Atlantic Ocean. This work underlined the role of the ocean in transmitting the signals of abrupt climate change.

Abrupt changes are often regional, for example, severe droughts lasting for many years have changed civilizations, and have occurred during the last 10 kyr of stable warm climate. This result has altered the notion of a stable climate during warm epochs, as previously suggested by the polar ice cores. The emerging picture of an unstable oceanatmosphere system has opened the debate of whether human interference through greenhouse gases and aerosols could trigger such events. Palaeoclimate reconstructions cited in the FAR were based on various data, including pollen records, insect and animal remains, oxygen isotopes and other geological data from lake varves, loess, ocean sediments, ice cores and glacier termini.

These records provided estimates of climate variability on time scales up to millions of years. A climate proxy is a local quantitative record that is interpreted as a climate variable using a transfer function that is based on physical principles and recently observed correlations between the two records.

The combination of instrumental and proxy data began in the 1960s with the investigation of the influence of climate on the proxy data, including tree rings, corals and ice cores. Phenological and historical data are also a valuable source of climatic reconstruction for the period before instrumental records became available. Such documentary data also need calibration against instrumental data to extend and reconstruct the instrumental record. With the development of multi-proxy reconstructions, the climate data were extended not only from local to global, but also from instrumental data to patterns of climate variability. Most of these reconstructions were at single sites and only loose efforts had been made to consolidate records.

Mann et al. made a notable advance in the use of proxy data by ensuring that the dating of different records lined up. Thus, the true spatial patterns of temperature variability and change could be derived, and estimates of NH average surface temperatures were obtained. The Working Group I WGI FAR noted that past climates could provide analogues. Fifteen years of research since that assessment has identified a range of variations and instabilities in the climate system that occurred during the last 2 Myr of glacial-interglacial cycles and in the super-warm period of 50 Ma. These past climates do not appear to be analogues of the immediate future, yet they do reveal a wide range of climate processes that need to be understood when projecting 21st century climate change.

SOLAR VARIABILITY AND THE TOTAL SOLAR IRRADIANCE

Measurement of the absolute value of total solar irradiance is difficult from the Earth's surface because of the need to correct for the influence of the atmosphere. Langley attempted to minimise the atmospheric effects by taking measurements from high on Mt. Whitney in California, and to estimate the correction for atmospheric effects by taking measurements at several times of day, for example, with the solar radiation having passed through different atmospheric pathlengths. Between 1902 and 1957, Charles Abbot and a number of other scientists around the globe made thousands of measurements of TSI from mountain sites. Values ranged from 1,322 to 1,465 W m-2, which encompasses the current estimate of 1,365 W m-2. Foukal et al. deduced from Abbot's daily observations that higher values of TSI were associated with more solar faculae.

In 1978, the Nimbus-7 satellite was launched with a cavity radiometre and provided evidence of variations in TSI. Additional observations were made with an active cavity radiometre on the Solar Maximum Mission, launched in

1980. Both of these missions showed that the passage of sunspots and faculae across the Sun's disk influenced TSI. At the maximum of the 11-year solar activity cycle, the TSI is larger by about 0.1 per cent than at the minimum. The observation that TSI is highest when sunspots are at their maximum is the opposite of Langley's hypothesis. As early as 1910, Abbot believed that he had detected a downward trend in TSI that coincided with a general cooling of climate. The solar cycle variation in irradiance corresponds to an 11-year cycle in radiative forcing which varies by about 0.2 W m-2.

There is increasingly reliable evidence of its influence on atmospheric temperatures and circulations, particularly in the higher atmosphere. Calculations with three-dimensional models suggest that the changes in solar radiation could cause surface temperature changes of the order of a few tenths of a degree celsius. For the time before satellite measurements became available, the solar radiation variations can be inferred from cosmogenic isotopes and from the sunspot number. Naked-eye observations of sunspots date back to ancient times, but it was only after the invention of the telescope in 1607 that it became possible to routinely monitor the number, size and position of these 'stains' on the surface of the Sun. Throughout the 17th and 18th centuries, numerous observers noted the variable concentrations and ephemeral nature of sunspots, but very few sightings were reported between 1672 and 1699.

This period of low solar activity, now known as the Maunder Minimum, occurred during the climate period now commonly referred to as the Little

Ice Age. There is no exact agreement as to which dates mark the beginning and end of the Little Ice Age, but from about 1350 to about 1850 is one reasonable estimate.' During the latter part of the 18th century, Wilhelm Herschel noted the presence not only of sunspots but of bright patches, now referred to as faculae, and of granulations on the solar surface. He believed that when these indicators of activity were more numerous, solar emissions of light and heat were greater and could affect the weather on Earth. Heinrich Schwabe published his discovery of a '10-year cycle' in sunspot numbers. Samuel Langley compared the brightness of sunspots with that of the surrounding photosphere. He concluded that they would block the emission of radiation and estimated that at sunspot cycle maximum the Sun would be about 0.1 per cent less bright than at the minimum of the cycle, and that the Earth would be 0.1°C to 0.3°C cooler.

These satellite data have been used in combination with the historically recorded sunspot number, records of cosmogenic isotopes, and the characteristics of other Sun-like stars to estimate the solar radiation over the last 1,000 years. These data sets indicated quasi-periodic changes in solar radiation of 0.24 to 0.30 per cent on the centennial time scale. These values have recently been re-assessed. The TAR states that the changes in solar irradiance are not the major cause of the temperature changes in the second half of the 20th century unless those changes can induce unknown large feedbacks in the climate system. The effects of galactic cosmic rays on the atmosphere and those due to shifts in the solar spectrum towards the ultraviolet range, at times of high solar activity, are largely unknown. The latter may produce changes in tropospheric circulation via changes in static stability resulting from the interaction of the increased UV radiation with stratospheric ozone. More research to investigate the effects of solar behaviour on climate is needed before the magnitude of solar effects on climate can be stated with certainty.

BIOGEOCHEMISTRY AND RADIATIVE FORCING

The modern scientific understanding of the complex and interconnected roles of greenhouse gases and aerosols in climate change has undergone rapid evolution over the last two decades. While the concepts were recognised and outlined in the 1970s, the publication of generally accepted quantitative results coincides with, and was driven in part by, the questions asked by the IPCC beginning in 1988. Thus, it is instructive to view the evolution of this topic as it has been treated in the successive IPCC reports. The WGI FAR codified the key physical and biogeochemical processes in the Earth system that relate a changing climate to atmospheric composition, chemistry,

Climate Change

the carbon cycle and natural ecosystems. The science of the time, as summarised in the FAR, made a clear case for anthropogenic interference with the climate system.

In terms of greenhouse agents, the main conclusions from the WGI FAR Policymakers Summary are still valid today:

- 'Emissions resulting from human activities are substantially increasing the atmospheric concentrations of the greenhouse gases: CO_2, CH_4, CFCs, N_2O';
- 'Some gases are potentially more effective';
- Feedbacks between the carbon cycle, ecosystems and atmospheric greenhouse gases in a warmer world will affect CO_2 abundances; and
- GWPs provide a metric for comparing the climatic impact of different greenhouse gases, one that integrates both the radiative influence and biogeochemical cycles.

The climatic importance of tropospheric ozone, sulphate aerosols and atmospheric chemical feedbacks were proposed by scientists at the time and noted in the assessment. For example, early global chemical modelling results argued that global tropospheric ozone, a greenhouse gas, was controlled by emissions of the highly reactive gases nitrogen oxides, carbon monoxide and non-methane hydrocarbons. In terms of sulphate aerosols, both the direct radiative effects and the indirect effects on clouds were acknowledged, but the importance of carbonaceous aerosols from fossil fuel and biomass combustion was not recognised.

The concept of radiative forcing as the radiative imbalance in the climate system at the top of the atmosphere caused by the addition of a greenhouse gas was established at the time and summarised in the WGI FAR. Agents of RF included the direct greenhouse gases, solar radiation, aerosols and the Earth's surface albedo. What was new and only briefly mentioned was that 'many gases produce indirect effects on the global radiative forcing'. The innovative global modelling work of Derwent showed that emissions of the reactive but non-greenhouse gases–NOx, CO and NMHCs–altered atmospheric chemistry and thus changed the abundance of other greenhouse gases. Indirect GWPs for NOx, CO and VOCs were proposed.

The projected chemical feedbacks were limited to short-lived increases in tropospheric ozone. By 1990, it was clear that the RF from tropospheric ozone had increased over the 20th century and stratospheric ozone had decreased since 1980 but the associated RFs were not evaluated in the assessments. Neither was the effect of anthropogenic sulphate aerosols, except to note in the FAR that 'it is conceivable that this radiative forcing has been of a comparable magnitude, but of opposite sign, to the greenhouse forcing earlier in the century'. Reflecting in general the community's concerns about this relatively new measure of climate forcing, RF bar charts appear only in the underlying FAR stages, but not in the FAR Summary. Only the long-lived greenhouse gases are shown, although sulphate aerosols direct effect in the future is noted with a question mark.

The cases for more complex chemical and aerosol effects were becoming clear, but the scientific community was unable at the time to reach general agreement on the existence, scale and magnitude of these indirect effects. Nevertheless, these early discoveries drove the research agendas in the early 1990s. The widespread development and application of global chemistrytransport models had just begun with international workshops. In the Supplementary Report to the FAR, the indirect chemical effects of CO, NOx and VOC were reaffirmed, and the feedback effect of CH_4 on the tropospheric hydroxyl radical was noted. Aerosolclimate interactions still focused on sulphates, and the assessment of their direct RF for the NH was now somewhat quantitative as compared to the FAR. Stratospheric ozone depletion was noted as causing a significant and negative RF, but not quantified. Ecosystems research at this time was identifying the responses to climate change and CO_2 increases, as well as altered CH_4 and N_2O fluxes from natural systems; however, in terms of a community assessment it remained qualitative.

By 1994, with work on SAR progressing, the Special Report on Radiative Forcing reported significant breakthroughs in a set of stages limited to assessment of the carbon cycle, atmospheric chemistry, aerosols and RF. The carbon budget for the 1980s was analysed not only from bottomup emissions estimates, but also from a top-down approach including carbon isotopes. A first carbon cycle assessment was performed through an international model and analysis workshop examining terrestrial and oceanic uptake to better quantify the relationship between CO_2 emissions and the resulting increase in atmospheric abundance. Similarly, expanded analyses of the global budgets of trace gases and aerosols from both natural and anthropogenic sources highlighted the rapid expansion of biogeochemical research.

The first RF bar chart appears, comparing all the major components of RF change from the pre-industrial period to the present. Anthropogenic soot aerosol, with a positive RF, was not in the 1995 Special Report but was added to the SAR. In terms of atmospheric chemistry, the first open-invitation modelling study for the IPCC recruited 21 atmospheric chemistry models to participate in a controlled study of photochemistry and chemical feedbacks. These studies demonstrated a robust consensus about some indirect effects, such as the CH4 impact on atmospheric chemistry, but great uncertainty about others, such as the prediction of tropospheric ozone changes.

The model studies plus the theory of chemical feedbacks in the CH4-CO- OH system firmly established that the atmospheric lifetime of a perturbation of CH4 emissions was about 50 per cent greater than reported in the FAR. There was still no consensus on quantifying the past or future changes in tropospheric ozone or OH. In the early 1990s, research on aerosols as climate forcing agents expanded. Based on new research, the range of climaterelevant aerosols was extended for the first time beyond sulphates to include nitrates, organics, soot, mineral dust and sea salt.

Quantitative estimates of sulphate aerosol indirect effects on cloud properties and hence RF were sufficiently well established to be included in assessments, and carbonaceous aerosols from biomass burning were recognised as being comparable in importance to sulphate. Ranges are given in the special report for direct sulphate RF and biomass-burning aerosols.

The aerosol indirect RF was estimated to be about equal to the direct RF, but with larger uncertainty. The injection of stratospheric aerosols from the eruption of Mt. Pinatubo was noted as the first modern test of a known radiative forcing, and indeed one climate model accurately predicted the temperature response. In the one-year interval between the special report and the SAR, the scientific understanding of aerosols grew. The direct anthropogenic aerosol forcing was reduced to -0.5 W m^{-2}. The RF bar chart was now broken into aerosol components with a separate range for indirect effects. Throughout the 1990s, there were concerted research programmes in the USA and EU to evaluate the global environmental impacts of aviation. Several national assessments culminated in the IPCC Special Report on Aviation and the Global Atmosphere which assessed the impacts on climate and global air quality.

An open invitation for atmospheric model participation resulted in community participation and a consensus on many of the environmental impacts of aviation. The direct RF of sulphate and of soot aerosols was likewise quantified along with that of contrails, but the impact on cirrus clouds that are sometimes generated downwind of contrails was not. The assessment re- affirmed that RF was a first-order metric for the global mean surface temperature response, but noted that it was inadequate for regional climate change, especially in view of the largely regional forcing from aerosols and tropospheric ozone. By the end of the 1990s, research on atmospheric composition and climate forcing had made many important advances. The TAR was able to provide a more quantitative evaluation in some areas. For example, a large, open-invitation modelling workshop was held for both aerosols and tropospheric ozone-OH chemistry.

This workshop brought together as collaborating authors most of the international scientific community involved in developing and testing global models of atmospheric composition. In terms of atmospheric chemistry, a strong consensus was reached for the first time that science could predict the changes in tropospheric ozone in response to scenarios for CH4 and the indirect greenhouse gases and that a quantitative GWP for CO could be reported. Further, combining these models with observational analysis, an estimate of the change in tropospheric ozone since the pre-industrial era–with uncertainties–was reported. The aerosol workshop made similar advances in evaluating the impact of different aerosol types. There were many different representations of uncertainty in the TAR, and the consensus RF bar chart did not generate a total RF or uncertainties for use in the subsequent IPCC Synthesis Report.

CLIMATE CHANGE MITIGATION

Climate change mitigation is action to decrease the intensity of radiative forcing in order to reduce the potential effects of global warming. Mitigation is distinguished from adaptation to global warming, which involves acting to tolerate the effects of global warming. Most often, climate change mitigation scenarios involve reductions in the concentrations of greenhouse gases, either by reducing their sources or by increasing their sinks.

Scientific consensus on global warming, together with the precautionary principle and the fear of abrupt climate change is leading to increased effort to develop new technologies and sciences and carefully manage others in an attempt to mitigate global warming. Most means of mitigation appear effective only for preventing further warming, not at reversing existing warming. The Stern Review identifies several ways of mitigating climate change. These include reducing demand for emissions-intensive goods and services, increasing efficiency gains, increasing use and development of low-carbon technologies, and reducing fossil fuel emissions.

The energy policy of the European Union has set a target of limiting the global temperature rise to 2 °C compared to preindustrial levels, of which 0.8 °C has already taken place and another 0.5–0.7 °C is already committed. The 2 °C rise is typically associated in climate models with a carbon dioxide-equivalent concentration of 400–500 ppm by volume; the current level of carbon dioxide alone is 383 ppm by volume, and rising at 2 ppm annually.

Hence, to avoid a very likely breach of the 2 °C target, CO2 levels would have to be stabilised very soon; this is generally regarded as unlikely, based on current programmes in place to date. The importance of change is showed by the fact that world economic energy efficiency is presently improving at only half the rate of world economic growth.

CLIMATE CHANGE AND ECOSYSTEMS

Unchecked global warming could affect most terrestrial ecoregions. Increasing global temperature means that ecosystems will change; some species are being forced out of their habitats because of changing conditions, while others are flourishing. Secondary effects of global warming, such as lessened snow cover, rising sea levels, and weather changes, may influence not only human activities but also the ecosystem.

For the IPCC Fourth Assessment Report, experts assessed the literature on the impacts of climate change on ecosystems. Rosenzweig et al. concluded that over the last three decades, human-induced warming had likely had a discernable influence on many physical and biological systems. Schneider et al. concluded, with very high confidence, that regional temperature trends had already affected species and ecosystems around the world. With high confidence, they concluded that climate change would result in the extinction of many species and a reduction in the diversity of ecosystems.

- *Terrestrial ecosystems and biodiversity*: With a warming of 3°C, relative to 1990 levels, it is likely that global terrestrial vegetation would become a net source of carbon. With high confidence, Schneider et al. concluded that a global mean temperature increase of around 4°C by 2100 would lead to major extinctions around the globe.
- *Marine ecosystems and biodiversity*: With very high confidence, Schneider et al. concluded that a warming of 2°C above 1990 levels would result in mass mortality of coral reefs globally.
- *Freshwater ecosystems*: Above about a 4°C increase in global mean temperature by 2100, Schneider et al. concluded, with high confidence, that many freshwater species would become extinct.

Studying the association between Earth climate and extinctions over the past 520 million years, scientists from the University of York write, "The global temperatures predicted for the coming centuries may trigger a new 'mass extinction event', where over 50 per cent of animal and plant species would be wiped out." Many of the species at risk are Arctic and Antarctic fauna such as polar bears and Emperor Penguins. In the Arctic, the waters of Hudson Bay are ice-free for three weeks longer than they were thirty years ago, affecting polar bears, which prefer to hunt on sea ice. Species that rely on cold weather conditions such as gyrfalcons, and Snowy Owls that prey on lemmings that use the cold winter to their advantage may be hit hard. Marine invertebrates enjoy peak growth at the temperatures they have adapted to, regardless of how cold these may be, and cold-blooded animals found at greater latitudes and altitudes generally grow faster to compensate for the short growing season.

Warmer-than-ideal conditions result in higher metabolism and consequent reductions in body size despite increased foraging, which in turn elevates the risk of predation. Indeed, even a slight increase in temperature during development impairs growth efficiency and survival rate in rainbow trout. Rising temperatures are beginning to have a noticeable impact on birds, and butterflies have shifted their ranges northward by 200 km in Europe and North America. Plants lag behind, and larger animals' migration is slowed down by cities and roads. In Britain, spring butterflies are appearing an average of 6 days earlier than two decades ago.

A 2002 article in Nature surveyed the scientific literature to find recent changes in range or seasonal behaviour by plant and animal species. Of species showing recent change, 4 out of 5 shifted their ranges towards the poles or higher altitudes, creating "refugee species". Frogs were breeding, flowers blossoming and birds migrating an average 2.3 days earlier each decade; butterflies, birds and plants moving towards the poles by 6.1 km per decade. A 2005 study concludes human activity is the cause of the temperature rise and resultant changing species behaviour, and links these effects with the predictions of climate models to provide validation for them. Scientists have observed that Antarctic hair grass is colonizing areas of Antarctica where previously their survival range was limited.

Mechanistic studies have documented extinctions due to recent climate change: McLaughlin et al. documented two populations of Bay checkerspot butterfly being threatened by precipitation change. Parmesan states, "Few studies have been conducted at a scale that encompasses an entire species" and McLaughlin et al. agreed "few mechanistic studies have linked extinctions to recent climate change." Daniel Botkin and other authors in one study believe that projected rates of extinction are overestimated.

Many species of freshwater and saltwater plants and animals are dependent on glacier-fed waters to ensure a cold water habitat that they have adapted to. Some species of freshwater fish need cold water to survive and to reproduce, and this is especially true with Salmon and Cutthroat trout. Reduced glacier run-off can lead to insufficient stream flow to allow these species to thrive. Ocean krill, a cornerstone species, prefer cold water and are the primary food source for aquatic mammals such as the Blue Whale. Alterations to the ocean currents, due to increased freshwater inputs from glacier melt, and the potential alterations to thermohaline circulation of the worlds oceans, may affect existing fisheries upon which humans depend as well.

The white lemuroid possum, only found in the mountain forests of northern Queensland, has been named as the first mammal species to be driven extinct by global warming. The White Possum has not been seen in over three years. These possums cannot survive extended temperatures over 30 °C which occurred in 2005. A final expedition to uncover any surviving White Possums is scheduled for 2009.

FORESTS

Pine forests in British Columbia have been devastated by a pine beetle infestation, which has expanded unhindered since 1998 at least in part due to the lack of severe winters since that time; a few days of extreme cold kill most mountain pine beetles and have kept outbreaks in the past naturally contained. The infestation, which has killed about half of the province's lodgepole pines is an order of magnitude larger than any previously recorded outbreak and passed via unusually strong winds in 2007 over the continental divide to Alberta. An epidemic also started, be it at a lower rate, in 1999 in Colourado, Wyoming, and Montana. The United States forest service predicts that between 2011 and 2013 virtually all 5 million acres of Colourado's lodgepole pine trees over five inches in diametre will be lost.

As the northern forests are a carbon sink, while dead forests are a major carbon source, the loss of such large areas of forest has a positive feedback on global warming. In the worst years, the carbon emission due to beetle infestation of forests in British Columbia alone approaches that of an average year of forest fires in all of Canada or five years worth of emissions from that country's transportation sources.

Besides the immediate ecological and economic impact, the huge dead forests provide a fire risk. Even many healthy forests appear to face an increased risk of forest fires because of warming climates. The 10-year average of boreal forest burned in North America, after several decades of around 10,000 km² has increased steadily since 1970 to more than 28,000 km² annually. Though this change may be due in part to changes in forest management practices, in the western U.S., since 1986, longer, warmer summers have resulted in a fourfold increase of major wildfires and a sixfold increase in the area of forest burned, compared to the period from 1970 to 1986. A similar increase in wildfire activity has been reported in Canada from 1920 to 1999.

Forest fires in Indonesia have dramatically increased since 1997 as well. These fires are often actively started to clear forest for agriculture. They can set fire to the large peat bogs in the region and the COreleased by these peat bog fires has been estimated, in an average year, to be 15 per cent of the quantity of CO produced by fossil fuel combustion.

MOUNTAINS

Mountains cover approximately 25 per cent of earth's surface and provide a home to more than one-tenth of global human population. Changes in global climate pose a number of potential risks to mountain habitats. Researchers expect that over time, climate change will affect mountain and lowland ecosystems, the frequency and intensity of forest fires, the diversity of wildlife, and the distribution of water.

Studies suggest that a warmer climate in the United States would cause lower-elevation habitats to expand into the higher alpine zone. Such a shift would encroach on the rare alpine meadows and other high-altitude habitats. High-elevation plants and animals have limited space available for new habitat as they move higher on the mountains in order to adapt to long-term changes in regional climate.

Changes in climate will also affect the depth of the mountains snowpacks and glaciers. Any changes in their seasonal melting can have powerful impacts on areas that rely on freshwater run-off from mountains. Rising temperature may cause snow to melt earlier and faster in the spring and shift the timing and distribution of run-off. These changes could affect the availability of freshwater for natural systems and human uses.

ECOLOGICAL PRODUCTIVITY

- A document by Smith and Hitz, it is reasonable to assume that the relationship between increased global mean temperature and ecosystem productivity is parabolic. Higher carbon dioxide concentrations will favourably affect plant growth and demand for water. Higher temperatures could initially be favourable for plant growth. Eventually, increased growth would peak then decline.
- The IPCC, a global average temperature increase exceeding 1.5–2.5°C would likely have a predominantly negative impact on ecosystem goods and services, *e.g.*, water and food supply.
- Research done by the Swiss Canopy Crane Project suggests that slow-growing trees only are stimulated in growth for a short period under higher CO_2 levels, while faster growing plants like liana benefit in the long term. In general, but especially in rainforests, this means that liana become the prevalent species; and because they decompose much

faster than trees their carbon content is more quickly returned to the atmosphere. Slow growing trees incorporate atmospheric carbon for decades.

TIME SCALE OF CLIMATIC CHANGE

The importance of considering different time scales when investigating climate change has already been identified. Climate varies on all time scales, in response to random and periodic forcing factors. Across all time periods from a few years to hundreds of millions of years there is a white (background) noise of random variations of the climate, caused by internal processes and associated feedback mechanisms, often referred to as stochastic or random mechanisms. Such randomness accounts for much of the climate variation, and owes its existence to the complex and chaotic behaviour of the climate system in responding to forcing. An essential corollary of the existence of random processes is that a large proportion of climate variation cannot be predicted.

Of far more relevance are the periodic forcing factors, for by understanding their mechanisms and the impacts they have on the global climate, it is possible to predict future climate change. How the climate system responds to periodic forcing factors, however, is often not clear. If it is assumed that the climate system responds in a linear fashion to periodic forcing, variations in climate will exhibit similar periodicity. If, however, the response of the system to forcing is strongly non-linear, the periodicities in the response will not necessarily be identical to the periodicities in the forcing factor(s). Frequently, the climate responds in a fashion intermediate between the two.

There are many climate forcing factors spanning an enormous range of periodicities. The longest, 200 to 500 million years, involves the passage of our Solar System through the galaxy, and the variations in galactic dust. These may be considered to be external forcing mechanisms. Other long time scale variations (10^6 to 10^8 years) include the non-radiative forcing mechanisms, such as continental drift, orogeny (mountain building) and isostasy (vertical movements in the Earth's crust affecting sea level). The response of the climate system to this combination of forcing factors itself depends upon the different response times of the various components of the system. The overall climatic response will then be determined by the interactions between the components. The atmosphere, surface snow and ice, and surface vegetation typically respond to climatic forcing over a period of hours to days. The surface ocean has a response time measured in years, whilst the deep ocean and mountain glaciers vary only over a period spanning hundreds of years. Large ice sheets advance and withdraw over thousands of years whilst parts of the geosphere (e.g. continental weathering of rocks) respond only to forcing periods lasting hundreds of thousands to millions of years.

The response of the climate system to episodes of forcing can be viewed as a form of resonance. When the time period of forcing matches most closely the response time of a particular system component, the climatic response will be greatest within that component.

Milankovitch forcing, for example, with periods of tens of thousands of years will be manifest in the response of the ice sheets, and the overall response of the climate system will be dominated by changes within the cryosphere. In addition, longer response times of certain components of the climate system modulate, through feedback processes, the short term responses. The response of the deep ocean to short term forcing (e.g. enhanced greenhouse effect, solar variations, for example, will tend to attenuate or smooth the response of the atmosphere.

Throughout the remainder, it should be recognised that a range of time scales applies to climate forcing mechanisms, radiative and non-radiative, external and internal, and to the response of the different components of the climate system.

CLIMATE SENSITIVITY

The concept of feedback is related to the climate sensitivity or climate stability. It is useful to have a measure of the strength of various feedback processes which determine the ultimate response of the climate system to any change in radiative forcing. In general terms, an initial change in temperature due to a change in radiative forcing, ÄTforcing, is modified by the complex

THE EFFECTS OF CLIMATE CHANGE ON ECOSYSTEMS

One way that scientists gain understanding of how global warming will affect ecosystems is to analyse the effects of past climate on paleoecosystems. A paleoecosystem is an ecosystem that existed in a former geologic time period. By relating vegetative cover to past climates, models can be developed. Once a reliable model is created, input variables such as CO_2 can be varied and the results analysed. An example of what the effect would be like on populations of Douglas fir in the northwestern United States if the CO_2 content in the atmosphere were double what it was before the Industrial Revolution was modeled by the U.S. Geological Survey and is shown in the illustration that follows. In a study conducted by the U.S. Global Change Research Programme in Washington, D.C., first in 2001, then updated in 2004, in trying to predict what the effects of future climate change would have on ecosystems, they concluded that "climate change has the potential to affect the structure, function, and regional distribution of ecosystems, and thereby affect the goods and services they provide." They based their conclusions on a modeling and analysis project they conducted called the Vegetation/Ecosystem Modeling and Analysis Project.

This project was used to generate future ecosystem scenarios for the conterminous United States based on model-simulated responses to both the Canadian and Hadley scenarios of climate change. The VEMAP was subsequently used in a validation exercise for a Dynamic Global Vegetation Model by Oregon State University and the U.S. Forest Service in 2008. Their MC1-DGVM was used as the input data in both the VEMAP and VINCERA. Their MC1 was run on both the VEMAP and VINCERA climate and soil input data to document how a change in the inputs can affect model outcome. The simulation results under the two sets of future climate scenarios were compared to see how different inputs can affect vegetation distribution and carbon budget projections.

The results indicated that "under all future scenarios, the interior west of the United States becomes woodier as warmer temperatures and available moisture allow trees to get established in grasslands areas. Concurrently, warmer and drier weather causes the eastern deciduous and mixed forests to shift to a more open canopy woodland or savannatype while boreal forests disappear almost entirely from the Great Lakes area by the end of the 21st century. While under VEMAP scenarios the model simulated large increases in carbon storage in a future woodier west, the drier VINCERA scenarios accounted for large carbon losses in the east and only moderate gains in the west. But under all future climate scenarios, the total area burned by wildfires increased." The similarities of the two models served to validate the VEMAP project.

The Hadley model was developed by the Hadley Centre for Climate Prediction and Research in England. Also referred to as the Met Office Hadley Centre for Climate Change, it is based at the headquarters of the Met Office in Exeter. It is the key institution in the United Kingdom for climate research. It is currently involved not only with understanding the physical, chemical, and biological processes within the climate system, but also with developing working models to explain current phenomena and to predict future climate change. It also monitors global and national climate variability and change and strives to determine the causes of the fluctuations. The Canadian climate model was developed by the Canadian Centre for Climate Modelling and Analysis. The CCCma is a division of the Climate Research Branch of Environment Canada based out of the University of Victoria, Victoria, British Columbia.

Its specific focus is on climate change and modeling. In the past nine years, the CCCma has produced three atmospheric and three atmospheric/oceanic general circulation models, making them one of the international leaders in climate change research. What they found was that over the next few decades climate change in the United States will most likely lead to increased plant productivity as a result of increasing levels of CO_2 in the atmosphere. There will also be an increase in terrestrial carbon storage for many parts of the country, especially the areas that become warmer and wetter.

The southeast will most likely see reduced productivity and, therefore, a decrease in carbon storage. By the end of the 21st century, many areas of the country will have experienced changes in the distribution of vegetation. Wetter areas will see the growth of more trees; drier areas will have drier soils that will cause forested areas to die off and be replaced by savanna/grassland ecosystems.

Modeling the vegetation evolution and adaptation is more difficult. The study focused on two time periods: 2025-2034 and 2090-2099. In the near term, biogeochemical changes are expected to dominate the ecological responses. Biogeochemical responses include changes based on the natural cycles of carbon, nutrients and water. The responses are affected by changing environmental conditions such as temperature, precipitation, solar radiation, soil texture, and atmospheric CO_2.

It is these natural cycles that affect carbon capture by plants with photosynthesis, soil nitrogen processes, and water transfer. These biogeochemical factors are what influence the production of vegetation. In the results from the near-term biogeochemical model, the scientists concluded that there would be an increase in CO_2. They estimate that currently the average carbon storage rate is 66/Tg/yr. The Hadley model predicted that carbon storage rates by 2025-2034 would increase to 117/Tg/yr. The Canadian model estimated CO_2 to increase 96 Tg/yr. The Canadian model projected that the southeastern ecosystems will lose carbon in the near term because they predict the climate there will become hot and dry. The biogeography models look at the changing landscape based on changes in CO_2, evapotranspiration, vegetation establishment, and competition between species, growth rates, and life cycle/mortality rates. In this model, scientists at both the Hadley and Canadian Centre for Climate Change agree that vegetation will be able to freely move from one location to another.

Changes in vegetation distribution will vary from region to region as follows:

- *Northeast*: Forests remain the dominant natural vegetation, but forest mixes will change. There will also be some increase in savannas and wetlands.
- *Southeast*: Forests remain the dominant ecosystem, but mixes change. Savannas and grasslands encroach on forests, especially towards the end of the 21st century. Drought and wildfires contribute significantly to forest destruction.
- *Midwest*: Forests remain the dominant land cover, but changes in species type occur. There will be a modest expansion of savannas and grasslands.
- *Great Plains*: Slight increase in woody vegetation.
- *West*: The areas of desert ecosystems shrink, and forest ecosystems grow.
- *Northwest*: Forested areas grow slightly.

A separate study conducted by the Canadian government predicts the following ecosystem changes as a result of changing climate over the next 100 years:

- *Coastlines*:
  - Flooding and erosion in coastal regions
  - Sea-level rise
- *Forests*:
  - Increase in pests
  - Increased levels of drought and wildfire
- *Plants and animals*:
  - Warmer temperatures could make water supplies more scarce, having a negative impact on plants and animals, not giving them time to adjust.
- *Crops*:
  - In some areas, warmer climate may allow a three- to five-week extension of the frost-free season, which could benefit commercial agriculture.
  - In other areas, drier soils and lack of water will have a negative impact on agricultural productivity.
- *Wells*:
  - The quality and quantity of drinking water may be threatened by increasing drought.
- Harsh weather:

- Winter storms, floods, drought, heat waves, and tornadoes could become more frequent and severe.
• *Fisheries*:
  - Populations and ranges of species sensitive to changes in water temperature will be negatively affected.
  - Salmon harvests will be lower in the Pacific.
  - Changes in ocean currents may have a negative impact on the fisheries in the Atlantic.
• *Lakes and rivers*:
  - Water levels will decline under the influence of drought, negatively affecting drinking water quality.
  - Use of lakes for transportation, recreation, and fishing, and the ability to generate electricity may be curtailed under droughtlike conditions.
  - Other areas that may have an increase in precipitation may experience flooding, rising sea levels, and severe storms.

In February 2005, the Met Office in Exeter, England, issued a report titled "Avoiding Dangerous Climate Change." The objective of the study was to determine what levels of CO_2 were considered the tipping point for dangerous climate change with harmful effects on ecosystems and what actions could be taken now to avoid such an outcome. In the report, then prime minister Tony Blair stated:"It is now plain that the emission of greenhouse gases... is causing global warming at a rate that is unsustainable." Environment Secretary Margaret Beckett stated,"The report's conclusions would be a shock to many people.

The thing that is perhaps not so familiar to members of the public... is this notion that we could come to a tipping point where change could be irreversible. We're not talking about it happening over five minutes, of course, maybe over a thousand years, but it's the irreversibility that I think brings it home to people." The report, published by the British government, says there is only a small chance of greenhouse gas emissions being kept below"dangerous" levels. It warns that the Greenland ice sheet could melt, causing sea levels to rise by 23 feet over the next 1,000 years. It also warns that developing countries will be the hardest hit. The report also states, based on the vulnerability of many of the world's ecosystems, that the European Union (EU) has adopted a target of preventing an increase in global temperature of more than 3.3°F (2°C). Some believe even that may be too high. The report states:"Above two degrees, the risks increase very substantially," with"potentially large numbers of extinctions" and"major increases in hunger and water shortage risks... particularly in developing countries." In order to meet their goals, British scientists have advised that CO_2 levels should be stabilized at 450 parts per million (ppm) or below. Currently the atmosphere contains 380 ppm. In response to this, the British government's chief scientific adviser, Sir David King, said that was unlikely to happen. He stated,"We're going to be at 400 ppm in 10 years' time. I predict that without any delight in saying it." Myles Allen, an expert on atmospheric physics at Oxford University, said that:"Assessing a 'safe level' of carbon dioxide in the atmosphere was 'a bit like asking a doctor what's a safe number of cigarettes to smoke per day.'"

The report does conclude, however, that there are technological options available to reduce CO_2 emissions that will need to be used. The study also concluded that the biggest obstacles involved with using these new technologies, along with renewable resources of energy and"clean coal," are the current economic investments and traditionally strong bond to the oil industry, cultural attitudes that oppose change, and simple lack of awareness by many people. Various conservation organizations currently involved in the battle against global warming, such as the Union of Concerned Scientists (UCS), the Defenders of Wildlife, and the World Wildlife Fund (WWF), also support these ideas.

IMPACT OF CLIMATE CHANGE ON AGRICULTURE

Despite technological advances, such as improved varieties, genetically modified organisms, and irrigation systems, weather is still a key factor in agricultural productivity, as well as soil properties and natural communities. The effect of climate on agriculture is related to variabilities in local climates rather than in global climate patterns. The Earth's average surface temperature has increased by 1 degree F in just over the last century. Consequently, agronomists consider any assessment has to be individually consider each local area. On the other hand, agricultural trade has grown in recent years, and now provides significant amounts of food, on a national level to major importing countries, as well as comfortable income to exporting ones. The international aspect of trade and security in terms of food implies the need to also consider the effects of climate change on a global scale.

A study published in Science suggests that, due to climate change, "southern Africa could lose more than 30 per cent of its main crop, maize, by 2030. In South Asia losses of many regional staples, such as rice, millet and maize could top 10 per cent". The 2001 IPCC Third Assessment Report concluded that the poorest countries would be hardest hit, with reductions in crop yields in most tropical and sub-tropical regions due to decreased water availability, and new or changed insect pest incidence. In Africa and

Latin America many rainfed crops are near their maximum temperature tolerance, so that yields are likely to fall sharply for even small climate changes; falls in agricultural productivity of up to 30 per cent over the 21st century are projected. Marine life and the fishing industry will also be severely affected in some places.

Climate change induced by increasing greenhouse gases is likely to affect crops differently from region to region. For example, average crop yield is expected to drop down to 50 per cent in Pakistan just as to the UKMO scenario whereas corn production in Europe is expected to grow up to 25 per cent in optimum hydrologic conditions.

More favourable effects on yield tend to depend to a large extent on realization of the potentially beneficial effects of carbon dioxide on crop growth and increase of efficiency in water use. Decrease in potential yields is likely to be caused by shortening of the growing period, decrease in water availability and poor vernalization.

In the long run, the climatic change could affect agriculture in several ways:

- Productivity, in terms of quantity and quality of crops
- Agricultural practices, through changes of water use and agricultural inputs such as herbicides, insecticides and fertilizers
- Environmental effects, in particular in relation of frequency and intensity of soil drainage, soil erosion, reduction of crop diversity
- Rural space, through the loss and gain of cultivated lands, land speculation, land renunciation, and hydraulic amenities.
- Adaptation, organisms may become more or less competitive, as well as humans may develop urgency to develop more competitive organisms, such as flood resistant or salt resistant varieties of rice.

They are large uncertainties to uncover, particularly because there is lack of information on many specific local regions, and include the uncertainties on magnitude of climate change, the effects of technological changes on productivity, global food demands, and the numerous possibilities of adaptation.

Most agronomists believe that agricultural production will be mostly affected by the severity and pace of climate change, not so much by gradual trends in climate. If change is gradual, there may be enough time for biota adjustment. Rapid climate change, however, could harm agriculture in many countries, especially those that are already suffering from rather poor soil and climate conditions, because there is less time for optimum natural selection and adaption.

PHYSICAL EVIDENCE FOR CLIMATIC CHANGE

Evidence for climatic change is taken from a variety of sources that can be used to reconstruct past climates. Reasonably complete global records of surface temperature are available beginning from the mid-late 19th century.

For earlier periods, most of the evidence is indirect—climatic changes are inferred from changes in proxies, indicators that reflect climate, such as vegetation, ice cores, dendrochronology, sea level change, and glacial geology.

HISTORICAL AND ARCHAEOLOGICAL EVIDENCE

Climate change in the recent past may be detected by corresponding changes in settlement and agricultural patterns. Archaeological evidence, oral history and historical documents can offer insights into past changes in the climate. Climate change effects have been linked to the collapse of various civilizations.

GLACIERS

Glaciers are considered among the most sensitive indicators of climate change, advancing when climate cools and retreating when climate warms. Glaciers grow and shrink, both contributing to natural variability and amplifying externally forced changes. A world glacier inventory has been compiled since the 1970s, initially based mainly on aerial photographs and maps but now relying more on satellites. This compilation tracks more than 100,000 glaciers covering a total area of approximately 240,000 km2, and preliminary estimates indicate that the remaining ice cover is around 445,000 km2. The World Glacier Monitoring Service collects data annually on glacier retreat and glacier mass balance From this data, glaciers worldwide have been found to be shrinking significantly, with strong glacier retreats in the 1940s, stable or growing conditions during the 1920s and 1970s, and again retreating from the mid 1980s to present.

The most significant climate processes since the middle to late Pliocene (approximately 3 million years ago) are the glacial and interglacial cycles. The present interglacial period (the Holocene) has lasted about 11,700 years. Shaped by orbital variations, responses such as the rise and fall of continental ice sheets and significant sea-level changes helped create the climate. Other changes, including Heinrich events, Dansgaard–Oeschger events and the Younger Dryas, however, show how glacial variations may also influence climate without the orbital forcing.

Glaciers leave behind moraines that contain a wealth of material—including organic matter, quartz, and potassium that may be dated—recording the periods in which a glacier advanced and retreated. Similarly, by tephrochronological techniques, the lack of glacier cover can be identified by the presence of soil or volcanic tephra horizons whose date of deposit may also be ascertained.

VEGETATION

A change in the type, distribution and coverage of vegetation may occur given a change in the climate; this much is obvious. In any given scenario, a mild change in climate may result in increased precipitation and warmth, resulting in improved plant growth and the subsequent sequestration of airborne CO_2. Larger, faster or more radical changes, however, may well result in vegetation stress, rapid plant loss and desertification in certain circumstances.

ICE CORES

Analysis of ice in a core drilled from a ice sheet such as the Antarctic ice sheet, can be used to show a link between temperature and global sea level variations. The air trapped in bubbles in the ice can also reveal the CO_2 variations of the atmosphere from the distant past, well before modern environmental influences. The study of these ice cores has been a significant indicator of the changes in CO_2 over many millennia, and continues to provide valuable information about the differences between ancient and modern atmospheric conditions.

DENDROCLIMATOLOGY

Dendroclimatology is the analysis of tree ring growth patterns to determine past climate variations. Wide and thick rings indicate a fertile, well-watered growing period, whilst thin, narrow rings indicate a time of lower rainfall and less-than-ideal growing conditions.

POLLEN ANALYSIS

Palynology is the study of contemporary and fossil palynomorphs, including pollen. Palynology is used to infer the geographical distribution of plant species, which vary under different climate conditions. Different groups of plants have pollen with distinctive shapes and surface textures, and since the outer surface of pollen is composed of a very resilient material, they resist decay. Changes in the type of pollen found in different layers of sediment in lakes, bogs, or river deltas indicate changes in plant communities. These changes are often a sign of a changing climate. As an example, palynological studies have been used to track changing vegetation patterns throughout the Quaternary glaciations and especially since the last glacial maximum.

INSECTS

Remains of beetles are common in freshwater and land sediments. Different species of beetles tend to be found under different climatic conditions. Given the extensive lineage of beetles whose genetic makeup has not altered significantly over the millennia, knowledge of the present climatic range of the different species, and the age of the sediments in which remains are found, past climatic conditions may be inferred.

SEA LEVEL CHANGE

Global sea level change for much of the last century has generally been estimated using tide gauge measurements collated over long periods of time to give a long-term average. More recently, altimetre measurements — in combination with accurately determined satellite orbits — have provided an improved measurement of global sea level change. To measure sea levels prior to instrumental measurements, scientists have dated coral reefs that grow near the surface of the ocean, coastal sediments, marine terraces, ooids in limestones, and nearshore archaeological remains. The predominant dating methods used are uranium series and radiocarbon, with cosmogenic radionuclides being sometimes used to date terraces that have experienced relative sea level fall.

MEASUREMENT OF CLIMATE ELEMENTS

MEASUREMENT OF TEMPERATURE

Many surface air temperature records extend back to the middle part of the last century. The measurement of the surface air temperature is essentially the same now as it was then, using a mercury-in-glass thermometre, which can be calibrated accurately and used down to -39°C, the freezing point of mercury. For lower temperatures, mercury is usually substituted by alcohol. Maximum and minimum temperatures measured during specified time periods, usually 24 hours, provide useful information for the construction and analysis of temperature time series.

Analysis involves the calculation of averages and variances of the data and the identification, using various statistical techniques, of periodic variations, persistence and trends in the time series. Observations of temperature from surface oceans are also collected in order to construct time series. In recent decades, much effort has also been directed towards the measurement of temperature at different levels in the atmosphere. There are now two methods of measuring temperatures at different altitudes: the conventional radiosonde network; and the microwave-sounding unit (MSU) on the TIROS-N series of satellites. The conventional network extends back to 1958 and the MSU data to 1979.

Temperature is a valuable climate element in climate observation because it directly provides a measure of the energy of the system under inspection. For example, a global average temperature † reveals information about the energy content of the Earth-atmosphere system. A higher temperature would indicate a larger energy content. Changes in temperature indicate changes in the energy balance, the causes of which were discussed. Variations in temperature are also subject to less variability than other elements such as rainfall and wind. In addition, statistical analysis of temperature time series is often less complex than that associated with other series. Perhaps most importantly of all, our own perception of the state of the climate are intimately linked to temperature.

MEASUREMENT OF RAINFALL

Rainfall is measured most simply by noting periodically how much has been collected in an exposed vessel since the time of the last observation. Care must be taken to avoid underestimating rainfall due to evaporation of the collected water and the effects of wind. Time series can be constructed and analysis performed in a similar manner to those of temperature.

The measurement of global rainfall offers an indirect or qualitative assessment of the energy of the Earth-atmosphere system. Increased heat storage will increase the rate of evaporation from the oceans (due to higher surface temperatures). In turn, the enhanced levels of water vapour in the atmosphere will intensify global precipitation. Rainfall is, however, subject to significant temporal and spatial variability, and the occurrence of extremes, and consequently, analysis of time series is more complex.

MEASUREMENT OF HUMIDITY

The amount of water vapour in the air can be described in at least 5 ways, in terms of:

- The water-vapour pressure;
- The relative humidity;
- The absolute humidity

- The mixing ratio
- The dewpoint.

A full account of these definitions may be found in Linacre. The standard instrument for measuring humidity is a psychrometre. This is a pair of identical vertical thermometres, one of which has the bulb kept wet by means of a muslin moistened by a wick dipped in water. Evaporation from the wetted bulb lowers its temperature below the air temperature (measured by the dry bulb thermometre).

The difference between the two measured values is used to calculate the air's water-vapour pressure, from which the other indices of humidity can be determined.

MEASUREMENT OF WIND

Wind is usually measured by a cup anemometre which rotates about a vertical axis perpendicular to the direction of the wind. The exposure of wind instruments is important any obstruction close by will affect measurements. Wind direction is also measured by means of a vane, accurately balanced about a truly vertical axis, so that it does not settle in any particular direction during calm conditions.

HOMOGENEITY

Non-climatic influences - inhomogeneities - can and do affect climatic observations. Any analyst using instrumental climate data must first assess the quality of the observations. A numerical series representing the variations of a climatological element is called homogeneous if the variations are caused only by fluctuations in weather and climate.

Leaving aside the misrecording of data, the most important causes of inhomogeneity are:
- Changes in instrument, exposure and measuring technique (for example, when more technologically advanced equipment is introduced);
- Changes in station location (i.e. when equipment is moved to a new site);
- Changes in observation times and methods used to calculate daily averages; and
- Changes in the station environment, particularly urbanisation (for example, the growth of a city around a pre-existing meteorological station).

When assessing the homogeneity of a climate record, there are three major sources of information: the variations evident in the record itself; the station history; and nearby station data. Visual examination and statistical analysis of the station record may reveal evidence of systematic changes or unusual behaviour which suggest inhomogeneity. For example, there may be a step- change in the mean, indicating a change in station location. A steady trend may indicate a progressive change in the station environment such as urbanisation. An extreme value may be due to a typing error.

Often these inhomogeneities may be difficult to detect and other evidence is needed to confirm their presence. One source of evidence is the station history, referred to as metadata. The station history should include details of any changes of location of the station, changes in instrumentation or changes in the timing and nature of observation. Very often, though, actual correction factors to observation data containing known inhomogeneities will be difficult to calculate, and in these cases, the record may have to be rejected.

The third approach to homogenisation involves empirical comparisons between stations close to each other. Over time scales of interest in climate change studies, nearby stations (i.e. within 10km of each other) should be subject to similar changes in monthly, seasonal and annual climate.

The only differences should be random. Any sign of systematic behaviour in the differences (e.g. a trend or step-change) would suggest the presence of inhomogeneities. In light of the foregoing discussion on homogeneity, careful attention has to be paid to eliminating sources of non-climatic error when constructing large-scale record, such as the global surface air temperature time series. This, and similar records including sea surface temperatures, rely on the collection of millions of individual observations from a huge network made up of thousands of climate stations.

A number of these have been reviewed by Jones*et al.*. The effects of urbanisation (the artificial warming associated with the growth of towns and cities around monitoring sites) were considered to be the greatest source of inhomogeneity, but even this, it was concluded, accounts for at most a 0.05°C warming (or 10 per cent of the observed warming) over the last 100 years. Conrad and Pollak (1962), Folland *et al.* (1990), and Jones *et al.* provide useful references investigating the problems of homogeneity and data reliability of instrumental records of climate data.

STATISTICAL ANALYSIS OF INSTRUMENTAL RECORDS

Once climate data has been collected and corrected for inhomogeneities, it will need to be analysed. The aim of any statistical analysis is to identify systematic behaviour in a data set and hence improve understanding of the processes at work to compliment the theory. Statistical analysis is a search for a signal in the data that can be distinguished from the background noise In climate change research that signal will be a periodic variation, a quasi-periodic variation, a trend, persistence or extreme events in the climate element under analysis.

PALAEOCLIMATE RECONSTRUCTION FROM PROXY DATA

Climate varies over different time scales, from years to hundreds of millions of years, and each periodicity is a manifestation of separate forcing mechanisms. In addition, different components of the climate system change and respond to forcing factors at different rates; in order to understand the role such components play in the evolution of climate it is necessary to have a record considerably longer than the time it takes for them to undergo significant changes.

Palaeoclimatology is the study of climate and climate change prior to the period of instrumental measurements. A geological chronology of Earth history is provided in the appendix, which will provide a useful time frame for the material discussed both in the remainder.

Instrumental records span only a tiny fraction ($<10\text{-}7$) of the Earth's climatic history and so provide a inadequate perspective on climatic variation and the evolution of the climate today and in the future. A longer perspective on climate variability can be obtained by the study of natural phenomena which are climate-dependent. Such phenomena provide a proxy record of the climate.

Many natural systems are dependent on climate, and from these it may be possible to derive palaeoclimatic information from them. By definition, such proxy records of climate all contain a climatic signal, but that signal may be weak and embedded in a great deal of random (climatic) background noise. In essence, the proxy material has acted as a filter, transforming climate conditions in the past into a relatively permanent record. Deciphering that record is often a complex business.

Principle sources of proxy data for palaeoclimatic reconstructions:

- Glaciological (Ice Cores)
  - Oxygen isotopes
  - Physical properties
  - Trace element and microparticle concentrations

Geological

A. *Sediments*:
1. Marine (ocean sediment cores)

i) Organic sediments (planktonic and benthic fossils) Oxygen isotopes Faunal and floral abundances Morphological variations
ii) Inorganic sediments Mineralogical composition and surface texture Distribution of terrigenous material Ice-rafted debris

Geochemistry

2. Terrestrial
    i) Periglacial features
    ii) Glacial deposits and erosional features
    iii) Glacio-eustatic features (shorelines)
    iv) Aeolian deposits (sand dunes)
    v) Lacustrine deposits/varves (lakes)

B. *Sedimentary Rocks*:
- Facies analysis
- Fossil/microfossil analysis
- Mineral analysis
- Isotope geochemistry

C. *Biological*:
- Tree rings (width, density, isotope analysis)
- Pollen (species, abundances)
- Insects

D. *Historical*:
- Meteorological records
- Parameteorological records (environmental indicators)
- Phenological records (biological indicators)

The major types of proxy climatic data available are listed. Each proxy material differs according to: a) its spatial coverage; b) the period to which it pertains; and c) its ability to resolve events accurately in time. Some proxy records, for example ocean floor sediments, reveal information about long periods of climatic change and evolution (10^7 years), with a low-frequency resolution (10^3 years). Others, such as tree rings are useful only during the last 10,000 years at most, but offer a high frequency (annual) resolution. The choice of proxy record (as with the choice of instrumental record) very much depends on what physical mechanism is under review. As noted, climate responds to different forcing mechanisms over different time scales, and proxy materials will contain necessary climatic information on these to a greater or lesser extent, depending on the three factors mentioned.

Other factors that have to be considered when using proxy records to reconstruct palaeoclimates include the continuity of the record and the accuracy to which it can be dated. Ocean sediments may provide continuous records for over 1 million years (Ma) but typically they are hard to date using existing techniques. Usually there is an uncertainty of +/− 5 per cent of the record's true age. Ice cores are easier to date but may miss layers from certain periods due to melting and wind erosion. Glacial deposits are highly episodic in nature, providing evidence only of discrete events in the past. Different proxy systems also have different levels of inertia with respect to climate, such that some systems may vary exactly in phase with climate whereas others lag behind by as much as several centuries.

Like climate construction from instrumental records, palaeoclimate reconstruction may be considered to proceed through a number of stages. The first stage is that of proxy data collection, followed by initial analysis and measurement. This results in primary data. The next stage involves the calibration of the data with modern climate records. In this, the uniformitarian principle is assumed, whereby contemporary climatic variations form a modern analogue for palaeoclimatic changes. It is important to be aware, however, of the possibility that palaeo-environmental conditions may not have modern analogues. The calibration may be only qualitative, involving subjective assessment, or it may be highly quantitative. The secondary data provide a record of past climatic variation. The third stage is the statistical analysis of this secondary data. The palaeoclimatic record is now statistically described and interpreted, providing a set of tertiary data.

Obviously not exhaustative. Bradley offers an excellent review of the various proxy methods and techniques employed to reconstruct Quaternary † palaeoclimatic change, whilst Frakes provides a useful commentary on the evidence for pre-Quaternary climates spanning most of geologic time. In the following, some of the more widely used proxy techniques will be reviewed. In all of the accounts, attention should be paid to the issues of reliability, dating, interpretation and meaning for all forms of climate reconstruction.

HISTORICAL RECORDS

Historical records have been used to reconstruct climates dating back several thousand years (i.e. for most of the Holocene). Historical proxy data can be grouped into three major categories. First, there are observations of weather phenomena *per se*, for example the frequency and timing of frosts or the occurrence of snowfall. Secondly, there are records of weather-dependent natural or environmental phenomena, termed parameteorological phenomena, such as droughts and floods. Finally, there are phenological records of weather- dependent biological phenomena, such as the flowering of trees, or the migration of birds.

Major sources of historical palaeoclimate information include: ancient inscriptions; annals and chronicles; government records; estate records; quasi-scientific writings; and fragmented early instrumental records.

There are a number of major difficulties in using this kind of information. First, it is necessary to determine exactly what the author meant in describing the particular event. How severe was the "severe" frost? What precisely does the term drought refer to? Content analysis - a standard historical technique - has been used to assess, in quantitative terms, the meaning of key climatological phrases in historical accounts. This approach involves assessment of the frequency of use made of certain words or phrases by a particular author. Nevertheless, the subjectivity of any personal account has to be carefully considered. Very often, the record was not kept for the benefit of the future reader, but to serve some independent purpose. During much of the dynastic era in China, for example, records of droughts and floods would be kept in order to gain tax exemptions at times of climatic adversity.

Secondly, the reliability of the account has to be assessed. It is necessary to determine whether or not the author had first-hand evidence of the meteorological events. Thirdly, it is necessary to date and interpret the information accurately. The representativeness of the account has to be assessed. Was the event a localised occurrence or can its spatial extent be defined by reference to other sources of information? What was the duration of the event? a day? a month? a year?

Finally, the data must, as with all proxy records, be calibrated against recent observations and cross-referenced with instrumental data This might be achieved by a construction of indices (e.g. the number of reports of frost per winter) which can be statistically related to analogous information derived from instrumental records.

ICE CORES

As snow and ice accumulates on polar and alpine ice caps and sheets, it lays down a record of the environmental conditions at the time of its formation. Information concerning these conditions can be extracted from ice and snow that has survived the summer melt by physical and chemical means. When melting does occur, the refreezing of meltwater can provide a measure of the summer conditions.

Palaeoclimate information has been obtained from ice cores by three main approaches.

These involve the analysis of:

- Stable isotopes of water;

- Dissolved and particulate matter in the firn † and ice; and
- The physical characteristics of the firn and ice, and of air bubbles trapped in the ice. Each approach has also provided a means of dating the ice at particular depths in the ice core.

STABLE ISOTOPE ANALYSIS

The basis for palaeoclimatic interpretations of variations in the stable isotope † content of water molecules is that the vapour pressure of $H_2^{16}O$ is higher than that of $H_2^{18}O$. Evaporation from a water body thus results in a vapour which is poorer in ^{18}O than the initial water; conversely, the remaining water is enriched in ^{18}O. During condensation, the lower vapour pressure of the $H_2^{18}O$ ensures that it passes more readily into the liquid state than water vapour made up of the lighter oxygen isotope. During the poleward transportation of water vapour, such isotope fractionation continues this preferential removal of the heavier isotope, leaving the water vapour increasingly depleted in $H_2^{18}O$. Because condensation is the result of cooling, the greater the fall in temperature, the lower the heavy isotope concentration will be. Isotope concentration in the condensate can thus be considered as a function of the temperature at which condensation occurs. Water from polar snow will thus be found to be most depleted in $H_2^{18}O$.

PHYSICAL AND CHEMICAL CHARACTERISTICS OF ICE CORES

The occurrence of melt features in the upper layers of ice cores are of particular palaeoclimatic significance. Such features include horizontal ice lenses and vertical ice glands which have resulted from the refreezing of percolating water. They can be identified by their deficiency in air bubbles.

The relative frequency of melt phenomena may be interpreted as an index of maximum summer temperatures or of summer warmth in general. Other physical features of ices cores which offer information to the palaeoclimatologist include variations in crystal size, air bubble fabric and crystallographic axis orientation.

significance is the atmospheric gas content, as the air pores are closed off during the densification of firn to ice. Considerable research effort has been devoted to the analysis of carbon dioxide concentrations of air bubbles trapped in ice cores. It will be seen that variations in atmospheric carbon dioxide may have played an important role in the glacial-interglacial climatic variations during the Quaternary.

Finally, variations of particulate matter, particularly calcium, aluminium, silicon and certain atmospheric aerosols can also be used as proxy palaeoclimatic indicators.

DATING ICE CORES

One of the biggest problems in any ice core study is determining the age-depth relationship. Many different approaches have been used and it is now clear that fairly accurate time scales can be developed for the last 10,000 years. Prior to that, there is increasing uncertainty about ice age. The problem lies with the fact that the age-depth is highly exponential, and ice flow models are often needed to determine the ages of the deepest sections of ice cores. For example, the upper 1000m of a core may represent 50,000 years, whilst the next 50m may span another 100,000 year time period, due to the severe compaction, deformation and flow of the ice sheet in question.

Radio isotope dating †, using 210Pb (lead), 32Si (silicon), 39Ar (argon) and 14C (carbon) have all been used with varying degrees of success, over different time scales, to determine the age of ice cores.

Certain components of ice cores may reveal quite distinct seasonal variations which enable annual layers to be identified, providing accurate time scales for the last few thousand years. Such seasonal variations may be found in ä18O values, trace elements and microparticles (Hammer et al., 1978).

Where characteristic layers of known ages can be detected, these provide valuable chronostratigraphic markers against which other dating methods can be verified. So-called reference horizons have resulted from major explosive volcanic eruptions. These inject large quantities of dust and gases (principally sulphur dioxide) into the atmosphere, where they are globally dispersed. The gases are converted into aerosols (principally of sulphuric acid) before being washed out in precipitation. Hence, after major eruptions, the acidity of snowfall increases significantly above background levels. By identifying highly acidic layers (using electrical conductivity) resulting from eruptions of known age, an excellent means of checking seasonally based chronologies is available.

DENDROCLIMATOLOGY

The study of the annual growth of trees and the consequent assembling of long, continuous chronologies for use in dating wood is called and climate is called dendroclimatology. Dendroclimatology offers a high resolution (annual) form of palaeoclimate reconstruction for most of the Holocene.

Climate Change

The annual growth of a tree is the net result of many complex and interrelated biochemical processes. Trees interact directly with the microenvironment of the leaf and the root surfaces. The fact that there exists a relationship between these extremely localised conditions and larger scale climatic parametres offers the potential for extracting some measure of the overall influence of climate on growth from year to year. Growth may be affected by many aspects of the microclimate: sunshine, precipitation, temperature, wind speed and humidity. Besides these, there are other non-climatic factors that may exert an influence, such as competition, defoliators and soil nutrient characteristics.

There are several subfields of dendroclimatology associated with the processing and interpretation of different tree-growth variables. Such variables include tree-ring width (the most commonly exploited information source, e.g. Briffa and Schweingruber,), densitometric parametres and chemical or isotopic variables.

A cross section of most temperate forest tree trunks † will reveal an alternation of lighter and darker bands, each of which is usually continuous around the tree circumference. These are seasonal growth increments produced by meristematic tissues in the cambium of the tree. Each seasonal increment consists of a couplet of earlywood (a light growth band from the early part of the growing season) and denser latewood (a dark band produced towards the end of the growing season), and collectively they make up the tree ring. The mean width of the tree ring is a function of many variables, including the tree species, tree age, soil nutrient availability, and a whole host of climatic factors.

The problem facing the dendroclimatologist is to extract whatever climatic signal † is available in the tree-ring data from the remaining background "noise".

Whenever tree growth is limited directly or indirectly by some climate variable, and that limitation can be quantified and dated, dendroclimatology can be used to reconstruct some information about past environmental conditions.

Only for trees growing near the extremities of their ecological amplitude †, where they may be subject to considerable climatic stresses, is it likely that climate will be a limiting factor. Commonly two types of climatic stress are recognised, moisture stress and temperature stress.

Trees growing in semi-arid regions are frequently limited by the availability of water, and dendroclimatic indicators primarily reflect this under growth limitations imposed by temperature; hence dendroclimatic indicators in such trees contain strong temperature signals.

Furthermore, climatic conditions prior to the growth period may precondition physiological processes within the tree and hence strongly influence subsequent growth. Consequently, strong serial correlation or autocorrelation may establish itself in the tree-ring record. A specific tree ring will contain information not just about the climate conditions of the growth year but information about the months and years preceding it.

Several assumptions underlie the production of quantitative dendroclimatic reconstructions. First, the physical and biological processes which link toady's environment with today's variations in tree growth must have been in operation in the past. This is the principle of uniformitarianism.

Second, the climate conditions which produce anomalies in tree-growth patterns in the past must have their analogue during the calibration period.

Third, climate is continuous over areas adjacent to the domain of the tree- ring network, enabling the development of a statistical transfer function relating growth in the network to climate variability inside and outside of it. Finally, it is assumed that the systematic relationship between climate as a limiting factor and the biological response can be approximated as a linear mathematical expression. Fritts provides a more exhaustive review of the assumptions involved in the use of dendroclimatology.

The general approach taken in dendroclimatic reconstruction is:

- To collect (sample) data from a set of trees (within a tree population) which have been selected on the basis that climate (e.g. temperature, moisture) should be a limiting factor;
- To assemble the data into a composite site chronology by cross-dating the individual series after the removal of age effects by standardization † . This master chronology increases the (climate) signal to (non-climate) noise ratio;
- To build up a network of site chronologies for a region;
- To identify statistical relationships between the chronology times series and the instrumental climate data for the recent period - the calibration period;
- To use these relationships to reconstruct climatic information from the earlier period covered by the tree-ring data, and;
- Finally, to test, or verify, the resulting reconstruction against independent data.

Bradley gives a full account of the methods (1 to 6 above) of palaeoclimate reconstruction from tree-ring analysis. This approach may be applied to all the climate-dependent tree-growth variables, specifically tree-ring width, but also wood density and isotopic measurements.

interannual variations contain a strong climatic signal. Density variations are particularly valuable in dendroclimatology because they to not change significantly with tree age, and the process of standardisation (removal of growth function) can therefore be avoided.

The use of isotopic measurements in dendroclimatology also avoids the need for a standardisation process. The basic premise of isotope dendroclimatology is that since 18O/16O and D/H (deuterium/hydrogen) variations in meteoric (atmospheric) waters are a function of temperature, tree growth which records such isotope variations should preserve a record of past temperature fluctuations.

Unfortunately, isotope fractionation effects within the tree, which are themselves temperature dependent, will create problems associated with this technique.

OCEAN SEDIMENTS

Billions of tonnes of sediment accumulate in the ocean basins every year, and this may be indicative of climate conditions near the ocean surface or on the adjacent continents. Sediments are composed of both biogenic (organic) and terrigenous (inorganic) materials.

The biogenic component includes the remnants of planktonic (surface ocean-dwelling) and benthic (deep-water- or sea floor-dwelling) organisms which provide a record of past climate and oceanic circulation. Such records may reveal information about past surface water temperatures, salinity, dissolved oxygen and nutrient availability. By contrast, the nature and abundance of terrigenous materials provides information about continental humidity-aridity variations, and the intensities and directions of winds. Ocean sediment records have been used to reconstruct palaeoclimate changes over a range of time scales, from thousands of years to millions and even tens of millions of years in the past.

CHANGES IN GLACIERS, ICE SHEETS, AND ICE SHELVES

Glaciers store water over relatively long timescales compared to rivers and lakes—hundreds to a few thousands of years. Ice sheets store water for even longer—tens of thousands of years. But the shorter glacier timescales are comparable to human timescales, so people notice how glaciers change, and these changes have obvious impacts on the human environment. Many river systems depend on glacier melt, which maintains the water supply through the summer.

As glaciers shrink, so does the frozen water supply they store. This is one of the reasons why it is important to measure how and understand why glaciers change over time.

affect global sea level. As terrestrial ice masses grow, sea level falls; and as masses shrink, sea level rises. At the last glacial maximum, about 18,000 years ago, the growth of ice sheets and glaciers caused sea level to lower by about 120 metres (395 feet). Most of that change was due to the formation of large ice sheets in northern North America and Europe, but mountain glaciers, too, had their role.

Glacier change may be quantified by a change in length, by a change in the surface area covered, or most rigourously by the glacier's mass balance, the difference between the volume of ice accumulated from snowfall and the volume of ice lost to melting or iceberg calving in a year. A glacier is "in balance" when gains and losses are equal. Because glaciers and ice sheets are remote and are difficult to observe in detail, the errors involved in estimating how they change are large. The best estimates indicate that while the Greenland and Antarctic ice sheets are near balance, glaciers and ice caps are shrinking worldwide, contributing as much as 0.25 millimetres (0.01 inch) per year to global sea-level rise.

MODERN GLACIER RETREAT

Mountain glaciers are always changing because they tend towards balance with ever-changing snowfall and melting rates. Some special glaciers change volume for other reasons . "Surge-type glaciers" experience rapid speed-ups that move ice downstream quickly, thinning the glacier, followed by quiescent periods in which the ice thickens once again. "Tidewater glaciers" end as a floating ice tongue in a lake or bay, where interaction with the lake bottom or seafloor, and iceberg calving events may complicate their flow.

Yet despite all the possible complications, glaciologists conclude that worldwide, glaciers are retreating rapidly as climate warms. For example, the glaciers of Mount Kilimanjaro, about which Hemingway wrote in 1938, are predicted to be gone by 2020. Ice first started to accumulate on that mountain nearly 12,000 years ago. Alaskan glaciers, measured with laser from a low-flying airplane, are thinning rapidly and contributing as much as 0.14 millimetres (0.005 inches) per year to sea level. The thinning rate has increased within the last decade. In Antarctica, ice-shelf collapse along its northward-reaching peninsula has also been linked to ongoing warming. Glacier retreat also is being monitored in the Himalayas, where increased melting can lead to flood risks for people living downstream. Around the world, glaciers are retreating in response to warming weather, and water resources and sea level will continue to be affected as they do.

ANTARCTIC ICE SHELF BREAKUPS

Iceberg calving is an efficient and normal process by which floating ice shelves maintain a steady mass balance over many years; but it also can be an important indicator of climate change. While temperatures in the interior of Antarctica have remained fairly steady since scientists arrived on the continent, the Antarctic Peninsula has warmed about 2.5°C over the last half of the twentieth century.

An ice shelf is a floating ice mass that is attached to the coast along at least one edge. The year 2002 brought notable events involving Antarctic ice shelves. Two massive icebergs broke off from the Ross Ice Shelf, a large sheet of glacial ice and snow extending from the Antarctic mainland into the southern Ross Sea. In a different event, an enormous piece broke away from the Larsen B Ice Shelf, located on the eastern side of the Antarctic Peninsula, in the largest single event in a 30-year series of ice-shelf retreats.

In the Larsen B Ice Shelf event, which reduced the shelf to a size not seen for about 12,000 years, a 3,275-square-kilometre (1,260-square-mile) sector disintegrated over the course of 35 days. The rapid break-up of the floating ice mass, which had survived thousands of years of climate variations, came at the end of one of the warmest summers on record.

When summertime temperatures rose close to and above freezing, snow on the surface of the Larsen Ice Shelf began to melt, eventually forming kilometres-wide ponds. That water is believed to have filled crevasses, which are deep, wedge-shaped cracks in the surface of the ice. Because water is denser than ice, it exerts extra pressure on the sides of a crevasse.

Four Moderate Resolution Imaging Spectroradiometre (MODIS) satellite images show the progressive breaking apart of a 3,275-square-kilometre (1,260- square-mile) northern section of the Larsen B Ice Shelf in Antarctica. The top left image was taken January 31, 2002, and the bottom right image was taken March 5, 2002. The snowy ice-shelf surface is clearly visible to the left of each frame, with the dark ocean waters to the right.

The mountains of the Antarctic Peninsula curve along the western (left) side of each frame. Meltwater ponds form patches and stripes on the surface of the shelf. A few small icebergs are floating in the dark ocean surface in pre-breakup images (top); the icebergs in the far lower right of the frames had calved 2 years earlier from an ice shelf farther south. After March 2002, long narrow icebergs and fragments too small to be resolved by the satellite were nearly all that remained of the Larsen B Ice Shelf.

Cause it to crack down to the bottom of the shelf. When the meltwater-enhanced fractures became widespread, they weakened the ice shelf, allowing it to quickly break apart.

Scientists cannot yet attribute specific ice events to global warming, due to lack of specific long-term data that would yield definitive causal relationships. Recent events must be viewed in the context of long-term, natural processes as well as the relatively short-term body of data collected since the Industrial Revolution.

RECORDS OF CLIMATE CHANGE

In addition to being affected by climate change, glaciers and ice sheets are archives of climate-change data. Each winter, new snow falls on the surface of the glacier. Whatever snow does not melt during the following summer will be buried by more snow the next winter. This "old snow" is called firn. The frozen water molecules and air trapped in the firn record the chemistry and temperature of the water vapour from which the snow formed and the atmosphere from which it fell.

As each year's firn layer is buried, that climate record is buried as well. The firn layers move down from the surface and are compressed as new layers pile on top. Eventually, the firn becomes dense, glacier ice. Most of the air has been squeezed out of the ice, but a few bubbles remain. When glaciologists drill down through the ice, they are drilling backward in time; consequently, ice cores drilled from glaciers and ice sheets reveal both regional and global climate trends.

The longest ice-core climate record comes from Vostok, a Russian scientific station in East Antarctica. The ice sheet is 3,623 metres (2.2 miles) thick at Vostok Station, and reaches back in time more than 400,000 years. The ice and the bubbles trapped within it have been studied by glaciologists from Russia, France, and the United States.

The ice core reveals that global atmospheric carbon dioxide (CO_2), methane (CH_4), and dust content rise and fall as global temperature and ice volume change. When the climate is warm, atmospheric CO_2 and CH_4 concentrations are large, and when climate is cool, those gases are less abundant. Atmospheric dust concentration changes in an opposite sense, indicating that warm, interglacial atmospheres are relatively moist, whereas in glacial times, the global atmosphere is relatively dry. These records also show that the present-day CO_2 level is larger than it was in the past warm times between glaciations.

Ice cores retrieved from mountain glaciers record both global and local changes. For example, ice cores from glaciers in Peru and Bolivia (in South

America) are being used to establish how tropical climate and the Amazonian rain forest change as Earth's climate warms and cools. These ice cores also contain long records of the El Niño Southern Oscillation. Tropical glaciers are melting fast as climate warms, and as that happens, their contribution to water resources decreases, their contribution to global sea level increases, and a valuable climate archive is lost. The first photographs made for World View of Global Warming, ten years ago, were of glaciers in Antarctica and Peru. This page shows a selection of the locations where we have documented glacier and ice cap retreat — a small set of images illustrating the overwhelming evidence from hundreds of glaciers and ice caps on every continent that global warming is severely affecting the water and glacial cycles of the planet. This is a profound change that unlike natural cycles like the Little Ice Age of the 16th and 17th centuries is proceeding very rapidly and appears tied to no natural cycles. The best correlation for this change to all but a handful of the 160,000 land glaciers and parts of the great ice caps is to rising atmospheric temperatures tied to increasing amounts of greenhouse gases.

The largest implication of this loss of glaciers is not the change in scenery, but the fact that the seasonal meltwater from glaciers, especially in Asia and South America, is the life support for billions of people. Large cities like Lima get much of their water from glaciers. In other parts of the world, glacier water keeps streams cool and full for salmon and other important wildlife. And as more and more water reaches the ocean, it is increasing sea level at a faster rate — which threatens every coastal city and shoreline. New scientific projections show at least a three foot (one metre) rise in ocean levels by the end of this century, part of which is also due to the expansion of warming sea water. This will inundate rice fields and estuaries that feed billions, and push into the heart of the worlds largest cities — and make each storm a threat of more devastating waves and surges.

The disintegrating face of the Müller Ice Shelf, Lallemand Fjord, Antarctic Peninsula, 67° South, April 2, 1999. This small shelf, fed by glaciers from the Loubet Coast, has been receding recently after growing over a 400-year cooling period. Like other receding ice shelves such as the larger Larsen, it may be a sensitive monitor of rising regional temperatures. The Larsen Ice Shelf lost a 1200 square mile section early in 2002, prompting some glaciologists to be concerned that even the giant Ross Ice Shelf could be at risk.

This mile-long ice cliff of Marr Ice Piedmont, Anvers Island, has receded about 500 metres since the mid 1960s. The cliff's previous position was to the left of the line of ice floating in the harbour and extended to the headland at the extreme upper left. The regional temperature has increased 5° C in winter over the past 50 years. This reduces seasonal icepack, disrupts growth of krill and changes conditions on penguin rookeries.

Photographer Gary Braasch holding a 1932 photo of Broggi glacier near Huascaran in the Peruvian Andes, while rephotographing this receding glacier in 1999. Glaciers everywhere in the world (with a very few exceptions) have been shrinking throughout the 20th Century, a prime signal of rapid global warming. Loss of tropical glaciers is particularly rapid. This glacier, previously photographed by the Austrian Hans Kinzl, receded about one kilometre in 67 years.

RADIATIVE FORCING AND CLIMATE SENSITIVITY

The radiative forcings from halocarbons and ozone to assess their roles in climate change. To make this comparison we have to assume that radiative forcings can be compared between different mechanisms and radiative forcing is related to the climate change issue of interest. The first assumption is equivalent to saying that the climate sensitivity is constant between different mechanisms and is implicit whenever direct and indirect radiative forcings are compared. However, several climate modelling studies have found that for many climate mechanisms ë varies with the latitude of the imposed forcing, and is higher for changes in the extratropics than for changes in the tropics. Additionally, several climate-modelling studies have compared the climate sensitivity for stratospheric ozone increases and generally found that global stratospheric ozone increases have a 20–80 per cent higher climate sensitivity than carbon dioxide.

This finding has been attributed to an additional positive feedback that results from an increase in stratospheric water vapour, which in turn arises from a warmer tropical tropopause. However, the latter studies all used idealized stratospheric ozone changes, and to date no study has performed similar analyses with realistic ozone changes. Climate models also typically have different responses to equivalent forcings from carbon dioxide and from other WMGHGs; these results indicate possible differences in climate sensitivity for CO2 and halocarbons. The two studies that have examined this directly reached contradictory conclusions. Hansen et al. found halocarbons to have about a 20 per cent larger climate sensitivity than carbon dioxide,

experiments. In contrast, Forster and Joshi found that, because halocarbons preferentially heat the tropical tropopause region rather than the surface, climate sensitivity for halocarbon changes was 6% smaller than for carbon dioxide changes.

The trade-off or partial cancellation between direct and indirect radiative forcing only occurs in the global mean: the latitudinal forcing patterns actually complement each other. Both stratospheric ozone depletion and increases in halocarbons realise more positive forcing in the tropics and more negative forcing at higher latitudes. Therefore, the variation in climate sensitivity with latitude of any applied forcing would likely mean that, even if the halocarbon and ozone forcing cancelled each other out in the global mean, an overall global cooling would result. Further, the patterns of surface temperature response may well be distinct. Radiative forcing estimates are measured as changes from pre-industrial times and are indicative of the equilibrium surface temperature change one might have expected since then. Radiative forcing is not necessarily indicative of patterns of temperature change, transient temperature changes or other metrics. From any time period, such as the present, future temperature changes will come from the combination of temperatures continuing to respond to past radiative forcing and any new changes to the radiative forcing. The rate of change of radiative forcing is potentially more important for evaluating short-term climate change and possibilities for mitigation than the total radiative forcing. It is therefore useful to examine the rate, as well as the total radiative forcing.

In summary, adopting radiative forcing and examining its rate of change are among the best methods for comparing the climate roles of carbon dioxide, ozone-depleting gases, their substitutes and ozone. However, radiative forcing gives an estimate only of the equilibrium globally averaged surface temperature response, and could lead to errors of around 50 per cent when comparing the predicted global mean temperature change from ozone and the halocarbons with that from carbon dioxide.

DIRECT RADIATIVE FORCING OF ODSS AND THEIR SUBSTITUTES

The direct radiative forcing from the ozone-depleting gases and their substitutes are relatively well known. The radiative forcing of the substitute gases is re-assessed in this report. The radiative forcings of the ODSs have individual uncertainties of about 10 per cent, and together they have contributed about 0.26 W m–2 to the total WMGHG radiative forcing since 1970. Past changes and a future scenario for the abundances of the ozone- depleting gases. A range of future scenarios for the substitute gases. In this part, we examine scenarios for future radiative forcing of the ODSs themselves. One possible future scenario is the A1B scenario from the Special Report on Emission Scenarios, which for the ODSs is consistent with the Ab scenario from WMO. However, because of their long lifetime in the atmosphere and their continued emission from the 'ODS bank', they could dominate the radiative forcing for the next four to five decades.

In this scenario HFCs dominate the radiative forcing by the end of the century, but this is only one example of several possible scenarios. As the production of ODSs has now essentially ceased, the principal issue for the emissions of ODSs concerns the treatment of the ODS bank. If emissions of ODSs from the bank continue at their current rate, ODS radiative forcing will still decrease. However, for at least the next two decades, the continued emission from the ODS bank is expected to have a comparable contribution to the total radiative forcing with that of HFC emissions. This calculation does not take into account the effect that ODSs have on the stratospheric ozone radiative forcing, which is discussed next.

INDIRECT RADIATIVE FORCING OF ODSS

IPCC gives the value for the radiative forcing of stratospheric ozone as -0.15 W m^{-2}, with a range of ± 0.1 W m^{-2}. This radiative forcing primarily arises from mid-and high-latitude ozone depletion in the lower stratosphere.

Past calculations of this radiative forcing have extensively employed observed trends in ozone. It is generally assumed that these observed ozone trends have been caused entirely by emissions of ODSs, so the negative radiative forcing of stratospheric ozone can be thought of as 'indirect'. For the ODSs it is possible to calculate EESC values and use them to scale the ozone radiative forcing.

These values have been calculated by assuming that EESC values above a 1979 background level give a forcing that scales with the -0.15 W m^{-2} best-estimate forcing from IPCC over 1979–1997. This calculation follows the approaches of Daniel et al. and Forster and Joshi. It results in a value of the indirect forcing that is purely from stratospheric ozone loss and ignores any possible changes to tropospheric chemistry and climate. Uncertainties in the indirect radiative forcing of stratospheric ozone arise from many different factors:

The quoted IPCC range of ± 0.1 W m^{-2} for the radiative forcing of stratospheric ozone is based on differing model results and arises primarily from the underlying uncertainty in the vertical distribution of the ozone trend relative to the tropopause. The future radiative forcing from ozone is in fact more uncertain than the range quoted in IPCC because there is a lack of confidence in the details of future ozone changes, particularly near the tropopause, and many different scenarios are possible. Although most of the ozone changes can be attributed to ODSs, a sizeable fraction of the NH changes could have other causes, especially those that occur close to the tropopause. This means that a small but possibly significant part attributable to ODSs. The simple linear scaling between EESC values, above a 1979 background level, and the stratospheric ozone radiative forcing is used for obtaining an approximation to the ozone radiative forcing time-series. It was

Climate Change

first proposed by Daniel et al. For EESC values below the 1979 level there is assumed to be no ozone depletion and no indirect radiative forcing.

NET RADIATIVE FORCING

Direct radiative forcing from halocarbons has risen sharply since 1980. However, ozone depletion has contributed a negative indirect forcing. Uncertainties in the magnitude of the indirect forcing are sufficiently large that the net radiative forcing since 1980 may have either increased or decreased. The best estimate is that the indirect forcing has largely offset the increase in the direct radiative forcing. This suggests that surface temperatures over the last few decades would have increased even more rapidly if stratospheric ozone depletion had not occurred. However, it is important to note that ozone depletion since 1980 is the result, in part, of ODS increases prior to 1980. The balance between direct and indirect radiative forcing varies substantially between different classes of ODSs. For the halons, the negative indirect forcing dominates their very small positive direct effect, causing their net forcing to be negative. In contrast, the CFCs and HCFCs have a net positive radiative forcing, despite their associated ozone loss. In that case, even though the direct radiative forcing of the halocarbons would decrease during the next few decades, the net forcing would actually increase because of ozone recovery. However, this increase in net forcing depends on which scenario is used for the ODS substitutes.

Effects of Ozone Depletion and Global Warming

(UV-B) radiation causes skin cancer, cataracts and immune suppression in both animals and humans. UV-B also damages plants including hardwood forests, and phytoplankton (an alga is a type of phytoplankton which is the building block of the oceanic food chain).

Effect on Infectious Diseases

Most infectious diseases are transmitted by insects and rodents. Transmitters of disease are called vectors. For example, mosquitoes transmit malaria, dengue and viral encephalitis (inflammation of the brain). Like other animals and plants, vectors are accustomed to certain climate conditions. If the climate becomes warmer, the mosquito will try to fly to new places where it can survive and expose more people to the disease. Changes in sea surface temperature and sea level can lead to higher incidence of water-borne infectious and toxin-related illnesses such as malaria (severe chills and fever), cholera (intestinal disease), dengue (characterized by severe pains in the joints and back), and leishmaniasis (skin ulcers). Human susceptibility to infections can be further compounded by malnutrition. UV-B radiation from ozone depletion damages both plants and animals. UV-B harms amphibian eggs, midge larvae and trout. Crops that are damaged will reduce food availability. UV-B also can damage mammalian immune systems which makes humans and other animals more susceptible to infectious diseases. Approximately 92 million people are expected to become refugees from global warming and climate change by 2100, not including any added from population growth.

OZONE DEPLETION

Ozone depletion describes two distinct, but related observations: a slow, steady decline of about 4 per cent per decade in the total volume of ozone in Earth's stratosphere (ozone layer) since the late 1970s, and a much larger, but seasonal, decrease in stratospheric ozone over Earth's polar regions during the same period. The latter phenomenon is commonly referred to as the ozone hole. In addition to this well-known stratospheric ozone depletion, there are also tropospheric ozone depletion events, which occur near the surface in Polar Regions during spring.

The detailed mechanism by which the polar ozone holes form is different from that for the mid-latitude thinning, but the most important process in both trends is catalytic destruction of ozone by atomic chlorine and bromine. The main source of these halogen atoms in the stratosphere is photodissociation of chlorofluorocarbon (CFC) compounds, commonly called freons, and of bromofluorocarbon compounds known as halons. These compounds are transported into the stratosphere after being emitted at the surface. Both ozone depletion mechanisms strengthened as emissions of CFCs and halons increased.

CFCs and other contributory substances are commonly referred to as ozone-depleting substances (ODS). Since the ozone layer prevents most harmful UVB wavelengths (270–315 nm) of ultraviolet light (UV light) from passing through the Earth's atmosphere, observed and projected decreases in ozone have generated worldwide concern leading to adoption of the Montreal Protocol banning the production of CFCs and halons as well as related ozone depleting chemicals such as carbon tetrachloride and trichloroethane.

It is suspected that a variety of biological consequences such as increases in skin cancer, damage to plants, and reduction of plankton populations in the ocean's photic zone may result from the increased UV exposure due to ozone depletion. Three forms (or allotropes) of oxygen are involved in the ozone-oxygen cycle: oxygen atoms (O or atomic oxygen), oxygen gas (O_2 or diatomic oxygen), and ozone gas (O_3 or triatomic oxygen). Ozone is formed in the stratosphere when oxygen molecules photodissociate after absorbing an ultraviolet photon whose wavelength is shorter than 240 nm.

This produces two oxygen atoms. The atomic oxygen then combines with O_2 to create O_3. Ozone molecules absorb UV light between 310 and 200 nm, following which ozone splits into a molecule of O_2 and an oxygen atom. The oxygen atom then joins up with an oxygen molecule to regenerate ozone. This is a continuing process which terminates when an oxygen atom "recombines" with an ozone molecule to make two O_2 molecules: $O + O_3 \rightarrow 2\ O_2$.

The overall amount of ozone in the stratosphere is determined by a balance between photochemical production and recombination. Ozone can be destroyed by a number of free radical catalysts, the most important of which are the hydroxyl radical (OH·), the nitric oxide radical (NO·) and atomic chlorine (Cl·) and bromine (Br·). All of these have both natural and anthropogenic (manmade) sources; at the present time, most of the OH· and NO· in the stratosphere is of natural origin, but human activity has dramatically increased the levels of chlorine and bromine.

These elements are found in certain stable organic compounds, especially chlorofluorocarbons (CFCs), which may find their way to the stratosphere without being destroyed in the troposphere due to their low reactivity. Once in the stratosphere, the Cl and Br atoms are liberated from the parent compounds by the action of ultraviolet light, e.g.

The Cl and Br atoms can then destroy ozone molecules through a variety of catalytic cycles. In the simplest example of such a cycle, a chlorine atom reacts with an ozone molecule, taking an oxygen atom with it (forming ClO) and leaving a normal oxygen molecule. The chlorine monoxide (i.e., the ClO) can react with a second molecule of ozone (i.e., O3) to yield another chlorine atom and two molecules of oxygen. The chemical shorthand for these gas- phase reactions is:

Cl + O3 '→ ClO + O2
ClO + O3 '→ Cl + 2 O2

The overall effect is a decrease in the amount of ozone. More complicated mechanisms have been discovered that lead to ozone destruction in the lower stratosphere as well.

A single chlorine atom would keep on destroying ozone (thus a catalyst) for up to two years (the time scale for transport back down to the troposphere) were it not for reactions that remove them from this cycle by forming reservoir species such as hydrogen chloride (HCl) and chlorine nitrate (ClONO2).

On a per atom basis, bromine is even more efficient than chlorine at destroying ozone, but there is much less bromine in the atmosphere at present.

As a result, both chlorine and bromine contribute significantly to the overall ozone depletion. Laboratory studies have shown that fluorine and iodine atoms participate in analogous catalytic cycles. However, in the Earth's stratosphere, fluorine atoms react rapidly with water and methane to form strongly-bound HF, while organic molecules which contain iodine react so rapidly in the lower atmosphere that they do not reach the stratosphere in significant quantities. Furthermore, a single chlorine atom is able to react with 100,000 ozone molecules. This fact plus the amount of chlorine released into the atmosphere by chlorofluorocarbons (CFCs) yearly demonstrates how dangerous CFCs are to the environment.

QUANTITATIVE UNDERSTANDING OF THE CHEMICAL OZONE LOSS PROCESS

New research on the breakdown of a key molecule in these ozone-depleting chemicals, dichlorine peroxide (Cl2O2), calls into question the completeness of present atmospheric models of polar ozone depletion. Specifically, chemists at NASA's Jet Propulsion Laboratory in Pasadena, California, found in 2007 that the temperatures, and the spectrum and intensity of radiation present in the stratosphere created conditions insufficient to allow the rate of chemical-breakdown required to release chlorine radicals in the volume necessary to explain observed rates of ozone depletion. Instead, laboratory tests, designed to be the most accurate reflection of stratospheric conditions to date, showed the decay of the crucial molecule almost a magnitude lower than previously thought.

PUBLIC POLICY IN RESPONSE TO THE OZONE HOLE

The full extent of the damage that CFCs have caused to the ozone layer is not known and will not be known for decades; however, marked decreases in column ozone have already been observed (as explained above). After a 1976 report by the U.S. National Academy of Sciences concluded that credible scientific evidence supported the ozone depletion hypothesis, a few countries, including the United States, Canada, Sweden, and Norway, moved to eliminate the use of CFCs in aerosol spray cans.

At the time this was widely regarded as a first step towards a more comprehensive regulation policy, but progress in this direction slowed in subsequent years, due to a combination of political factors (continued resistance from the halocarbon industry and a general change in attitude towards environmental regulation during the first two years of the Reagan administration) and scientific developments (subsequent National Academy assessments which indicated that the first estimates of the magnitude of ozone depletion had been overly large). The European Community rejected proposals to ban CFCs in aerosol sprays while even in the U.S., CFCs continued to be used as refrigerants and for cleaning circuit boards. Worldwide CFC production fell sharply after the U.S. aerosol ban, but by 1986 had returned nearly to its 1976 level. In 1980, DuPont closed down its research programme into halocarbon alternatives. The US Government's attitude began to change again in 1983, when William Ruckelshaus replaced Anne M. Burford as Administrator of the United States Environmental Protection Agency.

Under Ruckelshaus and his successor, Lee Thomas, the EPA pushed for an international approach to halocarbon regulations. In 1985 20 nations, including most of the major CFC producers, signed the Vienna Convention for the Protection of the Ozone Layer which established a framework for negotiating international regulations on ozone-depleting substances. That same year, the discovery of the Antarctic ozone hole was announced, causing a revival in public attention to the issue. In 1987, representatives from 43 nations signed the Montreal Protocol. Meanwhile, the halocarbon industry shifted its position and started supporting a protocol to limit CFC production.

The reasons for this were in part explained by "Dr. Mostafa Tolba, former head of the UN Environment Programme, who was quoted in the 30 June 1990 edition of The New Scientist, '...the chemical industry supported the Montreal Protocol in 1987 because it set up a worldwide schedule for phasing out CFCs, which [were] no longer protected by patents. This provided companies with an equal opportunity to market new, more profitable compounds.'" At Montreal, the participants agreed to freeze production of CFCs at 1986 levels and to reduce production by 50 per cent by 1999.

After a series of scientific expeditions to the Antarctic produced convincing evidence that the ozone hole was indeed caused by chlorine and bromine from manmade organohalogens, the Montreal Protocol was strengthened at a 1990 meeting in London. The participants agreed to phase out CFCs and halons entirely (aside from a very small amount marked for certain "essential" uses, such as asthma inhalers) by 2000. At a 1992 meeting in Copenhagen, the phase out date was moved up to 1996. To some extent, CFCs have been replaced by the less damaging hydro-chloro-fluoro-carbons (HCFCs), although concerns remain regarding HCFCs also. In some applications, hydro-fluoro-carbons (HFCs) have been used to replace CFCs. HFCs, which contain no chlorine or bromine, do not contribute at all to ozone depletion although they are potent greenhouse gases. The best known of these compounds is probably HFC-134a (R-134a), which in the United States has largely replaced CFC-12 (R-12) in automobile air conditioners.

In laboratory analytics (a former "essential" use) the ozone depleting substances can be replaced with various other solvents. *Ozone Diplomacy*, by Richard Benedick (Harvard University Press, 1991) gives a detailed account of the negotiation process that led to the Montreal Protocol. Pielke and Betsill provide an extensive review of early US government responses to the emerging science of ozone depletion by CFCs.

CURRENT EVENTS AND FUTURE PROSPECTS OF OZONE DEPLETION

Since the adoption and strengthening of the Montreal Protocol has led to reductions in the emissions of CFCs, atmospheric concentrations of the most significant compounds have been declining. These substances are being gradually removed from the atmosphere. By 2015, the Antarctic ozone hole would have reduced by only 1 million km^2 out of 25; complete recovery of the Antarctic ozone layer will not occur until the year 2050 or later. Work has suggested that a detectable (and statistically significant) recovery will not occur until around 2024, with ozone levels recovering to 1980 levels by around 2068.

The decrease in ozone-depleting chemicals has also been significantly affected by a decrease in bromine-containing chemicals. The data suggest that substantial natural sources exist for atmospheric methyl bromide (CH3Br). The 2004 ozone hole ended in November 2004, daily minimum stratospheric temperatures in the Antarctic lower stratosphere increased to levels that are too warm for the formation of polar stratospheric clouds (PSCs) about 2 to 3 weeks earlier than in most recent years.

The Arctic winter of 2005 was extremely cold in the stratosphere; PSCs were abundant over many high-latitude areas until dissipated by a big warming event, which started in the upper stratosphere during February and spread throughout the Arctic stratosphere in March. The size of the Arctic area of anomalously low total ozone in 2004-2005 was larger than in any year since 1997. The predominance of anomalously low total ozone values in the Arctic region in the winter of 2004-2005 is attributed to the very low stratospheric temperatures and meteorological conditions favourable for ozone destruction along with the continued presence of ozone destroying chemicals in the stratosphere.

A 2005 IPCC summary of ozone issues observed that observations and model calculations suggest that the global average amount of ozone depletion has now approximately stabilized. Although considerable variability in ozone is expected from year to year, including in polar regions where depletion is largest, the ozone layer is expected to begin to recover in coming decades due to declining ozone-depleting substance concentrations, assuming full compliance with the Montreal Protocol.

Temperatures during the Arctic winter of 2006 stayed fairly close to the long-term average until late January, with minimum readings frequently cold enough to produce PSCs. During the last week of January, however, a major warming event sent temperatures well above normal — much too warm to support PSCs. By the time temperatures dropped back to near normal in March, the seasonal norm was well above the PSC threshold.

Preliminary satellite instrument-generated ozone maps show seasonal ozone buildup slightly below the long-term means for the Northern Hemisphere as a whole, although some high ozone events have occurred.

During March 2006, the Arctic stratosphere poleward of 60 degrees North Latitude was free of anomalously low ozone areas except during the three-day period from 17 March to 19 when the total ozone cover fell below 300 DU over part of the North Atlantic region from Greenland to Scandinavia.

The area where total column ozone is less than 220 DU (the accepted definition of the boundary of the ozone hole) was relatively small until around 20 August 2006. Since then the ozone hole area increased rapidly, peaking at 29 million km^2 24 September. In October 2006, NASA reported that the year's ozone hole set a new area record with a daily average of 26 million km^2 between 7 September and 13 October 2006; total ozone thicknesses fell as low as 85 DU on 8 October. The two factors combined, 2006 sees the worst level of depletion in recorded ozone history. The depletion is attributed to the temperatures above the Antarctic reaching the lowest recording since comprehensive records began in 1979.

On October 2008 the Ecuadorian Space Agency published a report called HIPERION, a study of the last 28 years data from 10 satellites and dozens of ground instruments around the world among them their own, and found that the UV radiation reaching equatorial latitudes was far greater than expected, climbing in some very populated cities up to 24 UVI, the WHO UV Index standard considers 11 as an extreme index and a great risk to health. The report concluded that the ozone depletion around mid latitudes on the planet is already endangering large populations in this areas. Later, the CONIDA, the Peruvian Space Agency, made its own study, which found almost the same facts as the Ecuadorian study.

The Antarctic ozone hole is expected to continue for decades. Ozone concentrations in the lower stratosphere over Antarctica will increase by 5 per cent–10 per cent by 2020 and return to pre-1980 levels by about 2060–2075, 10–25 years later than predicted in earlier assessments. This is because of revised estimates of atmospheric concentrations of Ozone Depleting Substances — and a larger predicted future usage in developing countries. Another factor which may aggravate ozone depletion is the draw-down of nitrogen oxides from above the stratosphere due to changing wind patterns.

The basic physical and chemical processes that lead to the formation of an ozone layer in the earth's stratosphere were discovered by Sydney Chapman in 1930. These are discussed in the article Ozone-oxygen cycle — briefly, short- wavelength UV radiation splits an oxygen (O2) molecule into two oxygen (O) atoms, which then combine with other oxygen molecules to form ozone. Ozone is removed when an oxygen atom and an ozone molecule "recombine" to form two oxygen molecules, i.e. O + O3 '! 2O2.

In the 1950s, David Bates and Marcel Nicolet presented evidence that various free radicals, in particular hydroxyl (OH) and nitric oxide (NO),

Effects of Ozone Depletion and Global Warming

could These free radicals were known to be present in the stratosphere, and so were regarded as part of the natural balance – it was estimated that in their absence, the ozone layer would be about twice as thick as it currently is.

In 1970 Prof. Paul Crutzen pointed out that emissions of *nitrous* oxide (N_2O), a stable, long-lived gas produced by soil bacteria, from the earth's surface could affect the amount of *nitric* oxide (NO) in the stratosphere. Crutzen showed that nitrous oxide lives long enough to reach the stratosphere, where it is converted into NO.

Crutzen then noted that increasing use of fertilizers might have led to an increase in nitrous oxide emissions over the natural background, which would in turn result in an increase in the amount of NO in the stratosphere. Thus human activity could have an impact on the stratospheric ozone layer. In the following year, Crutzen and (independently) Harold Johnston suggested that NO emissions from supersonic aircraft, which fly in the lower stratosphere, could also deplete the ozone layer.

THE ROWLAND-MOLINA HYPOTHESIS

In 1974 Frank Sherwood Rowland, Chemistry Professor at the University of California at Irvine, and his postdoctoral associate Mario J. Molina suggested that long-lived organic halogen compounds, such as CFCs, might behave in a similar fashion as Crutzen had proposed for nitrous oxide. James Lovelock (most popularly known as the creator of the Gaia hypothesis) had recently discovered, during a cruise in the South Atlantic in 1971, that almost all of the CFC compounds manufactured since their invention in 1930 were still present in the atmosphere.

Molina and Rowland concluded that, like N_2O, the CFCs would reach the stratosphere where they would be dissociated by UV light, releasing Cl atoms. (A year earlier, Richard Stolarski and Ralph Cicerone at the University of Michigan had shown that Cl is even more efficient than NO at catalyzing the destruction of ozone. Similar conclusions were reached by Michael McElroy and Steven Wofsy at Harvard University. Neither group, however, had realized that CFC's were a potentially large source of stratospheric chlorine — instead, they had been investigating the possible effects of HCl emissions from the Space Shuttle, which are very much smaller.

The Rowland-Molina hypothesis was strongly disputed by representatives of the aerosol and halocarbon industries. The Chair of the Board of DuPont was quoted as saying that ozone depletion theory is "a science fiction tale...a load of rubbish...utter non-sense". Robert Abplanalp, the President of Precision Valve Corporation (and inventor of the first practical aerosol spray can valve), wrote to the Chancellor of UC Irvine to complain about Rowland's public statements (Roan, p 56.)

Nevertheless, within three years most of the basic assumptions made by Rowland and Molina were confirmed by laboratory measurements and by direct observation in the stratosphere. The concentrations of the source gases (CFCs and related compounds) and the chlorine reservoir species (HCl and ClONO2) were measured throughout the stratosphere, and demonstrated that CFCs were indeed the major source of stratospheric chlorine, and that nearly all of the CFCs emitted would eventually reach the stratosphere.

Even more convincing was the measurement, by James G. Anderson and collaborators, of chlorine monoxide (ClO) in the stratosphere. ClO is produced by the reaction of Cl with ozone — its observation thus demonstrated that Cl radicals not only were present in the stratosphere but also were actually involved in destroying ozone. McElroy and Wofsy extended the work of Rowland and Molina by showing that bromine atoms were even more effective catalysts for ozone loss than chlorine atoms and argued that the brominated organic compounds known as halons, widely used in fire extinguishers, were a potentially large source of stratospheric bromine.

In 1976 the U.S. National Academy of Sciences released a report which concluded that the ozone depletion hypothesis was strongly supported by the scientific evidence. Scientists calculated that if CFC production continued to increase at the going rate of 10 per cent per year until 1990 and then remain steady, CFCs would cause a global ozone loss of 5 to 7 per cent by 1995, and a 30 to 50 per cent loss by 2050.

In response the United States, Canada and Norway banned the use of CFCs in aerosol spray cans in 1978. However, subsequent research, summarized by the National Academy in reports issued between 1979 and 1984, appeared to show that the earlier estimates of global ozone loss had been too large. Crutzen, Molina, and Rowland were awarded the 1995 Nobel Prize in Chemistry for their work on stratospheric ozone.

THE OZONE HOLE

The discovery of the Antarctic "ozone hole" by British Antarctic Survey scientists Farman, Gardiner and Shanklin (announced in a paper in *Nature* in May 1985) came as a shock to the scientific community, because the observed decline in polar ozone was far larger than anyone had anticipated. Satellite measurements showing massive depletion of ozone around the south pole were becoming available at the same time.

However, these were initially rejected as unreasonable by data quality control algorithms (they were filtered out as errors since the values were unexpectedly low); the ozone hole was detected only in satellite data when the raw data was reprocessed following evidence of ozone depletion in *in situ* observations. When the software was rerun without the flags, the ozone hole was seen as far back as 1976.

Atmospheric Administration (NOAA), proposed that chemical reactions on polar stratospheric clouds (PSCs) in the cold Antarctic stratosphere caused a massive, though localized and seasonal, increase in the amount of chlorine present in active, ozone-destroying forms. The polar stratospheric clouds in Antarctica are only formed when there are very low temperatures, as low as - 80 degrees C, and early spring conditions.

In such conditions the ice crystals of the cloud provide a suitable surface for conversion of unreactive chlorine compounds into reactive chlorine compounds which can deplete ozone easily. Moreover the polar vortex formed over Antarctica is very tight and the reaction which occurs on the surface of the cloud crystals is far different from when it occurs in atmosphere. These conditions have led to ozone hole formation in Antarctica.

This hypothesis was decisively confirmed, first by laboratory measurements and subsequently by direct measurements, from the ground and from high-altitude airplanes, of very high concentrations of chlorine monoxide (ClO) in the Antarctic stratosphere. Alternative hypotheses, which had attributed the ozone hole to variations in solar UV radiation or to changes in atmospheric circulation patterns, were also tested and shown to be untenable.

Meanwhile, analysis of ozone measurements from the worldwide network of ground-based Dobson spectrophotometers led an international panel to conclude that the ozone layer was in fact being depleted, at all latitudes outside of the tropics. These trends were confirmed by satellite measurements. As a consequence, the major halocarbon producing nations agreed to phase out production of CFCs, halons, and related compounds, a process that was completed in 1996. Since 1981 the United Nations Environment Programme has sponsored a series of reports on scientific assessment of ozone depletion. The most recent is from 2007 where satellite measurements have shown the hole in the ozone layer is recovering and is now the smallest it has been for about a decade.

OZONE DEPLETION AND GLOBAL WARMING

Although they are often interlinked in the mass media, the connection between global warming and ozone depletion is not strong. There are five areas of linkage:

Radiative forcing from various greenhouse gases and other sources.
- The same CO2 radiative forcing that produces near-surface global warming is expected to cool the stratosphere. This cooling, in turn, is expected to produce a relative *increase* in polar ozone (O3) depletion and the frequency of ozone holes.
- Conversely, ozone depletion represents a radiative forcing of the climate system. There are two opposing effects: Reduced ozone causes the stratosphere to absorb less solar radiation, thus cooling the stratosphere while warming the troposphere; the resulting colder stratosphere emits less long-wave radiation downward, thus cooling the troposphere. Overall, the cooling dominates; the IPCC concludes that *"observed stratospheric O3 losses over the past two decades have caused a negative forcing of the surface-troposphere system"* of about "0.15 ± 0.10 watts per square meter (W/m^2).
- One of the strongest predictions of the greenhouse effect is that the stratosphere will cool. Although this cooling has been observed, it is not trivial to separate the effects of changes in the concentration of greenhouse gases and ozone depletion since both will lead to cooling. However, this can be done by numerical stratospheric modeling. Results from the National Oceanic and Atmospheric Administration's Geophysical Fluid Dynamics Laboratory show that above 20 km (12.4 miles), the greenhouse gases dominate the cooling.
- Ozone depleting chemicals are also greenhouse gases. The increases in concentrations of these chemicals have produced 0.34 ± 0.03 W/m^2 of radiative forcing, corresponding to about 14 per cent of the total radiative forcing from increases in the concentrations of well-mixed greenhouse gases.
- The long term modeling of the process, its measurement, study, design of theories and testing take decades to both document, gain wide acceptance, and ultimately become the dominant paradigm.

Several theories about the destruction of ozone, were hypothesized in the 1980s, published in the late 1990s, and are currently being proven. Dr Drew Schindell, and Dr Paul Newman, NASA Goddard, proposed a theory in the late 1990s, using a SGI Origin 2000 supercomputer, that modeled ozone destruction, accounted for 78 per cent of the ozone destroyed. Further refinement of that model, accounted for 89 per cent of the ozone destroyed, but pushed back the estimated recovery of the ozone hole from 75 years to 150 years. (An important part of that model is the lack of stratospheric flight due to depletion of fossil fuels.)

MISCONCEPTIONS ABOUT OZONE DEPLETION

A few of the more common misunderstandings about ozone depletion are addressed briefly here; more detailed discussions can be found in the ozone-depletion FAQ.

CFCS ARE "TOO HEAVY" TO REACH THE STRATOSPHERE

It is sometimes stated that since CFC molecules are much heavier than nitrogen or oxygen, they cannot reach the stratosphere in significant quantities. But atmospheric gases are not sorted by weight; the forces of wind (turbulence) are strong enough to fully intermix gases in the atmosphere. CFCs are heavier than air, but just like argon, krypton and other heavy gases with a long lifetime, they are uniformly distributed throughout the turbosphere and reach the upper atmosphere.

MAN-MADE CHLORINE IS INSIGNIFICANT COMPARED TO NATURAL SOURCES

Another objection occasionally voiced is that *It is generally agreed that natural sources of tropospheric chlorine (volcanoes, ocean spray, etc.) are four to five orders of magnitude larger than man-made sources.* While strictly true, *tropospheric* chlorine is irrelevant; it is *stratospheric* chlorine that affects ozone depletion. Chlorine from ocean spray is soluble and thus is washed out by rainfall before it reaches the stratosphere. CFCs, in contrast, are insoluble and long-lived, which allows them to reach the stratosphere.

Even in the lower atmosphere there is more chlorine present in the form of CFCs and related haloalkanes than there is in HCl from salt spray, and in the stratosphere halocarbons dominate overwhelmingly. Only one of these halocarbons, methyl chloride, has a predominantly natural source, and it is responsible for about 20 percent of the chlorine in the stratosphere; the remaining 80 per cent comes from manmade compounds.

Very large volcanic eruptions can inject HCl directly into the stratosphere, but direct measurements have shown that their contribution is small compared to that of chlorine from CFCs. A similar erroneous assertion is that soluble halogen compounds from the volcanic plume of Mount Erebus on Ross Island, Antarctica are a major contributor to the Antarctic ozone hole.

AN OZONE HOLE WAS FIRST OBSERVED IN 1956

G.M.B. Dobson mentioned that when springtime ozone levels over Halley Bay were first measured, he was surprised to find that they were ~320 DU, about 150 DU below spring levels, ~450 DU, in the Arctic. These, however, were the pre-ozone hole normal climatological values. What Dobson describes is essentially the *baseline* from which the ozone hole is measured: actual ozone hole values are in the 150–100 DU range.

The discrepancy between the Arctic and Antarctic noted by Dobson was primarily a matter of timing: during the Arctic spring ozone levels rose smoothly, peaking in April, whereas in the Antarctic they stayed approximately constant during early spring, rising abruptly in November when the polar vortex broke down. The behaviour seen in the Antarctic ozone hole is completely different. Instead of staying constant, early springtime ozone levels suddenly drop from their already low winter values, by as much as 50 per cent, and normal values are not reached again until December.

IF THE THEORY WERE CORRECT, THE OZONE HOLE SHOULD BE ABOVE THE SOURCES OF CFCS

CFCs are well mixed in the troposphere and the stratosphere. The reason the ozone hole occurs above Antarctica is not because there are more CFCs there but because the low temperatures allow polar stratospheric clouds to form. There have been anomalous discoveries of significant, serious, localized "holes" above other parts of the globe.

CONSEQUENCES OF OZONE LAYER DEPLETION

Since the ozone layer absorbs UVB ultraviolet light from the Sun, ozone layer depletion is expected to increase surface UVB levels, which could lead to damage, including increases in skin cancer. This was the reason for the Montreal Protocol.

Although decreases in stratospheric ozone are well-tied to CFCs and there are good theoretical reasons to believe that decreases in ozone will lead to increases in surface UVB, there is no direct observational evidence linking ozone depletion to higher incidence of skin cancer in human beings. This is partly due to the fact that UVA, which has also been implicated in some forms of skin cancer, is not absorbed by ozone, and it is nearly impossible to control statistics for lifestyle changes in the populace.

INCREASED UV

Ozone, while a minority constituent in the earth's atmosphere, is responsible for most of the absorption of UVB radiation. The amount of UVB radiation that penetrates through the ozone layer decreases exponentially with the slant-path thickness/density of the layer. Correspondingly, a decrease in atmospheric ozone is expected to give rise to significantly increased levels of UVB near the surface. Increases in surface UVB due to the ozone hole can

Effects of Ozone Depletion and Global Warming

be partially inferred by radiative transfer model calculations, but cannot be calculated from direct measurements because of the lack of reliable historical (pre-ozone-hole) surface UV data, although more recent surface UV observation measurement programmes exist (e.g. at Lauder, New Zealand).

Because it is this same UV radiation that creates ozone in the ozone layer from O2 (regular oxygen) in the first place, a reduction in stratospheric ozone would actually tend to increase photochemical production of ozone at lower levels (in the troposphere), although the overall observed trends in total column ozone still show a decrease, largely because ozone produced lower down has a naturally shorter photochemical lifetime, so it is destroyed before the concentrations could reach a level which would compensate for the ozone reduction higher up.

BIOLOGICAL EFFECTS OF INCREASED UV AND MICROWAVE RADIATION FROM A DEPLETED OZONE LAYER

The main public concern regarding the ozone hole has been the effects of surface UV on human health. So far, ozone depletion in most locations has been typically a few percent and, as noted above, no direct evidence of health damage is available in most latitudes. Were the high levels of depletion seen in the ozone hole ever to be common across the globe, the effects could be substantially more dramatic. As the ozone hole over Antarctica has in some instances grown so large as to reach southern parts of Australia and New Zealand, environmentalists have been concerned that the increase in surface UV could be significant.

EFFECTS OF OZONE LAYER DEPLETION ON HUMANS

UVB (the higher energy UV radiation absorbed by ozone) is generally accepted to be a contributory factor to skin cancer. In addition, increased surface UV leads to increased tropospheric ozone, which is a health risk to humans. The increased surface UV also represents an increase in the vitamin D synthetic capacity of the sunlight. The cancer preventive effects of vitamin D represent a possible beneficial effect of ozone depletion. In terms of health costs, the possible benefits of increased UV irradiance may outweigh the burden:

- *Basal and Squamous Cell Carcinomas:* The most common forms of skin cancer in humans, basal and squamous cell carcinomas, have been strongly linked to UVB exposure. The mechanism by which UVB induces these cancers is well understood — absorption of UVB radiation causes the pyrimidine bases in the DNA molecule to form dimers, resulting in transcription errors when the DNA replicates. These cancers are relatively mild and rarely fatal, although the treatment of squamous cell carcinoma sometimes requires extensive reconstructive surgery. By combining epidemiological data with results of animal studies, scientists have estimated that a one percent decrease in stratospheric ozone would increase the incidence of these cancers by 2 per cent.

- *Malignant Melanoma:* Another form of skin cancer, malignant melanoma, is much less common but far more dangerous, being lethal in about 15 per cent - 20 per cent of the cases diagnosed. The relationship between malignant melanoma and ultraviolet exposure is not yet well understood, but it appears that both UVB and UVA are involved. Experiments on fish suggest that 90 to 95 per cent of malignant melanomas may be due to UVA and visible radiation whereas experiments on opossums suggest a larger role for UVB. Because of this uncertainty, it is difficult to estimate the impact of ozone depletion on melanoma incidence. One study showed that a 10 per cent increase in UVB radiation was associated with a 19 per cent increase in melanomas for men and 16 per cent for women. A study of people in Punta Arenas, at the southern tip of Chile, showed a 56 per cent increase in melanoma and a 46 per cent increase in non-melanoma skin cancer over a period of seven years, along with decreased ozone and increased UVB levels.
- *Cortical Cataracts:* Studies are suggestive of an association between ocular cortical cataracts and UV-B exposure, using crude approximations of exposure and various cataract assessment techniques. A detailed assessment of ocular exposure to UV-B was carried out in a study on Chesapeake Bay Watermen, where increases in average annual ocular exposure were associated with increasing risk of cortical opacity. In this highly exposed group of predominantly white males, the evidence linking cortical opacities to sunlight exposure was the strongest to date. However, subsequent data from a population-based study in Beaver Dam, WI suggested the risk may be confined to men. In the Beaver Dam study, the exposures among women were lower than exposures among men, and no association was seen. Moreover, there were no data linking sunlight exposure to risk of cataract in African Americans, although other eye diseases have different prevalences among the different racial groups, and cortical opacity appears to be higher in African Americans compared with whites.
- *Increased Tropospheric Ozone:* Increased surface UV leads to increased tropospheric ozone. Ground-level ozone is generally recognized to be

a health risk, as ozone is toxic due to its strong oxidant properties. At this time, ozone at ground level is produced mainly by the action of UV radiation on combustion gases from vehicle exhausts.

EFFECTS ON CROPS

An increase of UV radiation would be expected to affect crops. A number of economically important species of plants, such as rice, depend on cyanobacteria residing on their roots for the retention of nitrogen. Cyanobacteria are sensitive to UV light and they would be affected by its increase.

EFFECTS ON PLANKTON

Research has shown a widespread extinction of plankton 2 million years ago that coincided with a nearby supernova. There is a difference in the orientation and motility of planktons when excess of UV rays reach earth. Researchers speculate that the extinction was caused by a significant weakening of the ozone layer at that time when the radiation from the supernova produced nitrogen oxides that catalyzed the destruction of ozone (plankton are particularly susceptible to effects of UV light, and are vitally important to marine food webs).

OBSERVATIONS ON OZONE LAYER DEPLETION

The most pronounced decrease in ozone has been in the lower stratosphere. However, the ozone hole is most usually measured not in terms of ozone concentrations at these levels (which are typically of a few parts per million) but by reduction in the total *column ozone*, above a point on the Earth's surface, which is normally expressed in Dobson units, abbreviated as "DU".

Marked decreases in column ozone in the Antarctic spring and early summer compared to the early 1970s and before have been observed using instruments such as the Total Ozone Mapping Spectrometer (TOMS).

Reductions of up to 70 per cent in the ozone column observed in the austral (southern hemispheric) spring over Antarctica and first reported in 1985 (Farman et al. 1985) are continuing. Through the 1990s, total column ozone in September and October have continued to be 40–50 per cent lower than pre- ozone-hole values. In the Arctic the amount lost is more variable year-to-year than in the Antarctic. The greatest declines, up to 30 per cent, are in the winter and spring, when the stratosphere is colder.

Reactions that take place on polar stratospheric clouds (PSCs) play an important role in enhancing ozone depletion. PSCs form more readily in the extreme cold of Antarctic stratosphere. This is why ozone holes first formed, and are deeper, over Antarctica. Early models failed to take PSCs into account and predicted a gradual global depletion, which is why the sudden Antarctic ozone hole was such a surprise to many scientists.

In middle latitudes it is preferable to speak of ozone depletion rather than holes. Declines are about 3 per cent below pre-1980 values for 35–60°N and about 6 per cent for 35–60°S. In the tropics, there are no significant trends. Ozone depletion also explains much of the observed reduction in stratospheric and upper tropospheric temperatures. The source of the warmth of the stratosphere is the absorption of UV radiation by ozone, hence reduced ozone leads to cooling. gases such as CO_2; however the ozone-induced cooling appears to be dominant. Predictions of ozone levels remain difficult. The World Meteorological Organization Global Ozone Research and Monitoring Project - Report No. 44 comes out strongly in favour for the Montreal Protocol, but notes that a UNEP 1994 Assessment overestimated ozone loss for the 1994–1997 period.

CHEMICALS IN THE ATMOSPHERE

CFCs in the Atmosphere

Chlorofluorocarbons (CFCs) were invented by Thomas Midgley in the 1920s. They were used in air conditioning/cooling units, as aerosol spray propellants prior to the 1980s, and in the cleaning processes of delicate electronic equipment. They also occur as by-products of some chemical processes. No significant natural sources have ever been identified for these compounds — their presence in the atmosphere is due almost entirely to human manufacture. As mentioned in the *ozone cycle overview* above, when such ozone-depleting chemicals reach the stratosphere, they are dissociated by ultraviolet light to release chlorine atoms.

The chlorine atoms act as a catalyst, and each can break down tens of thousands of ozone molecules before being removed from the stratosphere. Given the longevity of CFC molecules, recovery times are measured in decades. It is calculated that a CFC molecule takes an average of 15 years to go from the ground level up to the upper atmosphere, and it can stay there for about a century, destroying up to one hundred thousand ozone molecules during that time.

VERIFICATION OF OBSERVATIONS

Scientists have been increasingly able to attribute the observed ozone depletion to the increase of anthropogenic halogen compounds from CFCs by the use of complex chemistry transport models and their validation

Effects of Ozone Depletion and Global Warming

against observational data (e.g. SLIMCAT, CLaMS). These models work by combining satellite measurements of chemical concentrations and meteorological fields with chemical reaction rate constants obtained in lab experiments. They are able to identify not only the key chemical reactions but also the transport processes which bring CFC photolysis products into contact with ozone.

THE OZONE HOLE AND ITS CAUSES

The Antarctic ozone hole is an area of the Antarctic stratosphere in which the recent ozone levels have dropped to as low as 33 per cent of their pre-1975 values. The ozone hole occurs during the Antarctic spring, from September to early December, as strong westerly winds start to circulate around the continent and create an atmospheric container. Within this polar vortex, over 50 per cent of the lower stratospheric ozone is destroyed during the Antarctic spring. As explained above, the overall cause of ozone depletion is the presence of chlorine-containing source gases (primarily CFCs and related halocarbons).

1997 (Abnormally Cold Resulting in Increased Seasonal Depletion). In the presence of UV light, these gases dissociate, releasing chlorine atoms, which then go on to catalyze ozone destruction. The Cl-catalyzed ozone depletion can take place in the gas phase, but it is dramatically enhanced in the presence of polar stratospheric clouds (PSCs). These polar stratospheric clouds form during winter, in the extreme cold.

Polar winters are dark, consisting of 3 months without solar radiation (sunlight). Not only lack of sunlight contributes to a decrease in temperature but also the polar vortex traps and chills air. Temperatures hover around or below −80 °C.

These low temperatures form cloud particles and are composed of either nitric acid (Type I PSC) or ice (Type II PSC). Both types provide surfaces for chemical reactions that lead to ozone destruction. The photochemical processes involved are complex but well understood. The key observation is that, ordinarily, most of the chlorine in the stratosphere resides in stable "reservoir" compounds, primarily hydrochloric acid (HCl) and chlorine nitrate ($ClONO_2$).

During the Antarctic winter and spring, however, reactions on the surface of the polar stratospheric cloud particles convert these "reservoir" compounds into reactive free radicals (Cl and ClO). The clouds can also remove NO_2 from the atmosphere by converting it to nitric acid, which prevents the newly formed depletion is the reason why the Antarctic ozone depletion is greatest during spring.

During winter, even though PSCs are at their most abundant, there is no light over the pole to drive the chemical reactions. During the spring, however, the sun comes out, providing energy to drive photochemical reactions, and melt the polar stratospheric clouds, releasing the trapped compounds. Most of the ozone that is destroyed is in the lower stratosphere, in contrast to the much smaller ozone depletion through homogeneous gas phase reactions, which occurs primarily in the upper stratosphere. Warming temperatures near the end of spring break up the vortex around mid-December. As warm, ozone- rich air flows in from lower latitudes, the PSCs are destroyed, the ozone depletion process shuts down, and the ozone hole closes.

INTEREST IN OZONE LAYER DEPLETION

While the effect of the Antarctic ozone hole in decreasing the global ozone is relatively small, estimated at about 4 per cent per decade, the hole has generated a great deal of interest because:
- The decrease in the ozone layer was predicted in the early 1980s to be roughly 7 per cent over a 60 year period.
- The sudden recognition in 1985 that there was a substantial "hole" was widely reported in the press. The especially rapid ozone depletion in Antarctica had previously been dismissed as a measurement error.
- Many were worried that ozone holes might start to appear over other areas of the globe but to date the only other large-scale depletion is a smaller ozone "dimple" observed during the Arctic spring over the North Pole. Ozone at middle latitudes has declined, but by a much smaller extent (about 4–5 per cent decrease).
- If the conditions became more severe (cooler stratospheric temperatures, more stratospheric clouds, more active chlorine), then global ozone may decrease at a much greater pace. Standard global warming theory predicts that the stratosphere will cool.
- When the Antarctic ozone hole breaks up, the ozone-depleted air drifts out into nearby areas. Decreases in the ozone level of up to 10 per cent have been reported in New Zealand in the month following the break-up of the Antarctic ozone hole.

EFFECTS ON EARTH'S FOOD CHAIN

Ozone depletion and global warming have harmful effects on plants and animals. If allowed to continue, our food chain will be seriously disrupted. For example, phytoplantkton are tiny floating algae in the ocean which are the base of the marine food chain. In Antarctica, there has been upwards

Effects of Ozone Depletion and Global Warming

of 50 percent ozone depletion. This means that an unusually high amount of UV-B radiation has reached the Earth's surface in the Antarctic region. UV-B harms the productivity of phytoplankton, thereby reducing the available food for animals that feed on phytoplankton. Krill eat phytoplankton and penguins eat krill.

From a climate change perspective, phytoplankton normally absorbs a lot of carbon from the air. As phytoplankton dies from UV-B radiation, this carbon is no longer absorbed. This means that more carbon will be left in the atmosphere, contributing to more global warming. More global warming can increase ozone depletion, which kills more phytoplankton, and the process repeats itself.

"Total Environmental Impact" of Refrigerants

Six parameters define the "total environmental impact" of refrigerants on our environment.
- Ozone Depletion Potential (ODP)
- Global Warming Potential (GWP)
- Atmospheric Life
- Energy Use
- Equipment Emission Rate
- Refrigerant Charge

Approximately 70 percent of the world's electricity is generated by the burning of fossil fuels. For every additional kWh used, there are more greenhouse gas emissions generated by electric utility power plants. This is the indirect effect of global warming refrigerants. The environmental impact of even small changes in chiller energy use has an impact. There is a need to use refrigerants that minimize both ODP and GWP to address global environmental concerns. If the efficiency of every centrifugal chiller in the world were increased by only 0.08 kW/ton, power plant-generated greenhouse gas emissions would be reduced by literally billions of pounds. This is an amount equal to removing nearly two million cars from the road each year, or to planting nearly a half billion trees every year.

ACTION PLAN FOR ORGANIZATIONS USING REFRIGERANTS

Most refrigerants used in air conditioning and refrigeration contribute to global warming in addition to ozone depletion. Even the new non-ozone depleting alternative refrigerants add to the global warming problem. Businesses and organizations using refrigerants are encouraged to take action now to do what's right for both their organization and for the environment. Persons responsible for HVAC and/or environmental concerns in their company or organization need to take strong, immediate actions to reduce ozone depletion and global warming. Minimize emissions of ozone depleting (ODP) and direct global warming (GWP) refrigerants used by your organization. The most effective way to start reducing emissions is to implement a Refrigerant Management Plan and insure that your organization is complying with EPA's Section 608 Refrigerant Recycling Regulations. Refrigerants should be handled as a controlled substance.

A defined leak management programme, including leak monitoring and leak repair policies is essential. Provide Refrigerant Management and Regulations Compliance Training to your responsible employees. Organization's using refrigerants should review and implement industry best practices in refrigerant management to minimize emissions at every stage of refrigerant handling. Implement a process to keep current with regulatory changes to insure going compliance requirements are met. Reduce energy consumption of HVAC systems using refrigerants to reduce indirect global warming emissions. Conduct a building tune-up to reduce your heating, cooling, and electrical loads, and thus your overall energy consumption.

Replace older HVAC systems containing high ozone depleting and/or global warming refrigerants with newer, more efficient systems. Frequently, HVAC systems are oversized. Thus, you may want to consider right-sizing your existing system with a smaller, more energy-efficient one that matches the newly reduced loads. The operating costs of a high efficiency system may offer an attractive pay back of your investment. Evaluate HVAC systems on a total cost of ownership basis, rather than first cost.

MODELLING GLOBAL WARMING

Global average temperature has increased by around 0.6°C during the 20th century. At the same time, greenhouse gas concentrations have increased substantially. To assess whether the two are associated involves the detection of a human"fingerprint" on 20th century climate change. Detection of this man-made influence requires the use of computer model simulations of the likely climatic effects of changing the atmospheric composition, and the comparison of the results with observations. Computers model the climate by expressing the basic physical processes which control weather and climate as a series of mathematical equations. The climate however, is a very complex system, and supercomputers are needed for the task.

The most complex climate models are called general circulation models or GCMs because they model the circulation of the atmosphere. Coupled ocean- atmosphere GCMs also model the climatic influences of the oceans,

Effects of Ozone Depletion and Global Warming

which store so much energy, and the ocean-atmosphere interactions. GCMs can simulate both global average temperature changes and inter-regional differences. In addition, the climatic influence of both greenhouse gases and aerosols can be incorporated. By taking into account the global cooling potential of man-made aerosols, GCMs have simulated a rise in temperature close to the observed 0.6°C. This provides strong but not conclusive evidence that mankind has contributed towards global warming.

NITROUS OXIDE

Nitrous oxide (N2O) is a colourless, non-flammable gas with a sweetish odour, commonly known as "laughing gas", and sometimes used as an anaesthetic. Nitrous oxide is naturally produced by oceans and rainforests. Man-made sources of nitrous oxide include nylon and nitric acid production, the use of fertilisers in agriculture, cars with catalytic converters and the burning of organic matter. Nitrous oxide is broken down in the atmosphere by chemical reactions that involve sunlight.

Like carbon dioxide and methane, nitrous oxide is a greenhouse gas whose molecules absorb heat trying to escape to space. Nitrous oxide contributes to the Earth's natural greenhouse effect. Man-made emissions of nitrous oxide are helping to enhance the greenhouse effect. Since the beginning of the Industrial Revolution, atmospheric nitrous oxide concentration has increased by about 16 per cent, and has contributed 4 to 6 per cent to the enhancement of the greenhouse effect.

RAINFALL

With global warming and a rise in the global average surface temperature, increases in global rainfall and other forms of precipitation would be expected, due to the greater rates of evaporation of sea surface water. Unfortunately, no reliable estimates of evaporation increase exist. One problem is the effect of varying wind speed on evaporation rates, which may or may not be related to increases in temperature.

Several large-scale regional analyses of precipitation changes have been carried out. These have demonstrated that during the latter part of the 20th century precipitation has tended to increase in the mid-latitudes, for example in the former Soviet Union, but decrease in the Northern Hemisphere subtropics. A striking rainfall decrease occurred in the African Sahel north of the Sahara Desert, between the 1960s and 1980s. This dramatic desiccation has been linked to changes in ocean circulation and tropical Atlantic sea surface temperatures. Whether such changes are linked to global warming however, remains open to analysis.

The accuracy of other precipitation records should be treated with caution. Precipitation is more difficult to monitor than temperature due to its greater geographical variability. Other uncertainties in the data set may be due to the collection efficiency of rain gauges. Consequently, the compilation of a global precipitation record can prove to be very difficult and is perhaps unjustified.

SEA LEVEL

The global sea level has already risen by around between 10 to 25 centimetres during the last 100 years, at the rate of 1 to 2 millimetres per year. Measuring past and current changes in sea level, however, is extremely difficult. There are many potential sources of error and systematic bias, such as the uneven geographical distribution of measuring sites and the effect of the land itself as it rises and subsides.

It is likely that most of this rise in sea level has been due to the increase in global temperature over the last 100 years. Global warming should, on average, cause the oceans to warm and expand thus increasing sea level. Climate models indicate that about 25 per cent of the rise in sea level this century has been due to the thermal expansion of seawater. A second major cause of rising sea level is the melting of land-based ice caps. Presently, it is uncertain to what extent the melting of the Greenland and Antarctic ice caps has contributed to global sea level rise during the 20th century. indicate that the Earth's average surface temperature may increase by between 1.4 and 5.8oC during the 21st century. Global warming is expected to cause a further rise of between 9 and 88 centimetres by the year 2100, with a best estimate of 50 centimetres, if emissions of greenhouse gases remain uncontrolled. This expected rate of change (an average of 5 cm per decade) is significantly faster than that experienced over the last 100 years.

Forecasting sea level rise, however, involves many uncertainties. While most scientists believe that man-made greenhouse gas emissions are changing the climate and particularly the speed, of this change. Global warming is the main potential impact of greenhouse gas emissions, but other aspects of the climate besides temperature may also change. For example, some studies suggest that changes in precipitation will increase snow accumulation in Antarctica, which may help to moderate the net sea level rise. Another complication is that the sea level would not rise by the same amount all over the globe due to the effects of the Earth's rotation, local coastline variations, changes in major ocean currents, regional land subsidence and emergence, and differences in tidal patterns and sea water density.

Effects of Ozone Depletion and Global Warming

Nevertheless, some areas of Antarctica have warmed by 2.5oC during the past 50 years, a rate of warming 5 times faster than for the Earth as a whole. Whilst scientists believe this to reflect mostly regional changes in climate, the recent summertime disintegration of the Larsen Ice Shelf has renewed speculation that climatic changes in the polar regions have the potential to cause severe impacts via a rise in global sea level over the next 100 to 200 years.

TEMPERATURE

An analysis of combined land and ocean temperatures records indicates that during the last decade globally averaged surface temperatures have been higher than in any decade since the mid 18th century.

In fact, when records of tree rings are used to reconstruct temperatures back a thousand years, the 1990s may be the warmest period in a millennium. During the 20th century, a global temperature increase of about 0.6oC years has been observed.

Temperatures however, have increased rather differently in the two Hemispheres. A rapid increase in the Northern Hemisphere temperature during the 1920s and 1930s contrasts with a more gradual increase in the Southern Hemisphere. Both Hemispheres had relatively stable temperatures from the 1940s to the 1970s, although there is some evidence of cooling in the Northern Hemisphere during the 1960s. Since the 1960s in the Southern Hemisphere but after 1975 in the Northern Hemisphere, temperatures have risen sharply.

Whilst globally averaged records of temperature offer a means of assessing climate change, it is important to recognise that they represent an over-simplification. Significant latitudinal and regional differences in the extent and timing of warming exist. In addition, winter temperatures and night-time minimums may have risen more than summer temperatures and daytime maximums.

Temperature changes higher up in the atmosphere are central to the problem of human-induced global warming, because climate models predict that temperature changes with elevated concentrations of greenhouse gases will have a characteristic profile through the layers of the atmosphere, with warming up to about 6 km, and cooling at higher altitudes within the stratosphere.

The cooler stratospheric temperatures are an expected consequence of the increased trapping of heat from the Earth in the lowest part of the atmosphere. Records of stratospheric temperature confirm that cooling at higher altitudes has indeed occurred, particularly since 1980. The altitude at which warming changes to cooling however, is higher than that predicted by computer models.

TREES

A change in global climate would be accompanied by shifts in climatic zones, thereby altering the suitability of a region for the growth of distinctive species. Trees in particular have long reproductive cycles, and many species may not be able to respond to the climatic changes quickly enough. A shift in climatic zones not only affects the vegetation but also affects the incidence of tree pests such as insects and diseases. These pests have less difficulty in migrating with their climatic zones than vegetation and may damage tree species with lower immunity.

As well as the effects of temperature and precipitation variations, and changes to weather patterns, forest growth may also respond to increased atmospheric concentrations of carbon dioxide. Studies with immature forest plantations suggest that an increase in atmospheric carbon dioxide would be beneficial to tree growth. The elevated carbon dioxide concentrations enhance photosynthesis rates with increased utilisation of carbon dioxide. This is called the carbon fertilisation effect. As a consequence of carbon fertilisation, water use efficiency may also increase. Increase in growth rates however, would vary enormously within ecosystems and between species.

In general, it is expected that the negative impacts of climate change on forests will have a greater impact than any positive effect due to an increase in growth rates as a result of elevated atmospheric carbon dioxide concentrations. With unmitigated emissions of greenhouse gases, substantial dieback of tropical forests and tropical grasslands is predicted to occur by the emissions are reduced enabling atmospheric carbon dioxide concentrations to stabilise at 550 ppm (double the pre-industrial level), this loss would be substantially reduced, even by the 2230s. Considerable growth of forests is predicted to occur in North America, northern Asia and China.

As well as the effects on forests themselves, climate change is expected to influence societies and economies dependent upon forestry. Forest products make up the third most valuable international commodity after oil and gas. Trade is expected to increase in the 21st century along with demand, particularly in the developing countries.

Global warming may well affect the development of such developing economies, particularly if current rates of deforestation remain unchecked and the unsustainable management of forests continues. As a consequence, societies dependent upon the income, food and shelter their forests provide them may well face increasing stresses due to crop failures, soil nutrient

Effects of Ozone Depletion and Global Warming

depletion and the effects of extreme weather events in the years to come.

WATER

Global warming will lead to an intensification of the global water or hydrological cycle through increases in surface temperature and rates of evaporation, and in some regions, increases in precipitation. Changes in the total amount of precipitation and its frequency and intensity directly affect the magnitude and timing of run-off and the intensity of floods and droughts. Such changes will have significant impacts on regional water resources.

It is not certain how individual water catchment areas will respond to changing evaporation rates and precipitation. It is likely however, that currently dry regions will be more sensitive to changes in climate. Relatively small changes in temperature and precipitation could cause relatively large changes in run-off. Arid and semi-arid regions will therefore be particularly sensitive to reduced rainfall and to increased evaporation.

An increase in the duration of dry spells will not necessarily lead to an increased likelihood of low river flows and groundwater levels, since increases in precipitation may be experienced during other seasons.

More probably, increased rainfall will lead to an increased likelihood of river flooding. Changes in seasonal patterns of rainfall may affect the regional distribution of both ground and surface water supplies.

Hydrological regimes in high latitude or mountain areas are often determined by winter snowfall and spring snowmelt. Most climate models predict that global warming will reduce the amount of precipitation falling as snow in these regions, increasing the rate of water run-off and enhancing the likelihood of flooding. Climatic effects on tropical hydrological regimes are harder to predict. In the mid-latitudes, including the UK, wintertime soil moisture is expected to increase whilst summertime soil moisture may decrease. There will however, be regional variations.

Freshwater ecosystems, including lakes, streams and non-coastal wetlands will be influenced by changes to the hydrological cycle as a result of global warming. These influences will interact with other man-made changes in land use, waste disposal and water extraction. In general, freshwater organisms will tend to move towards higher latitudes as temperatures increase, whilst extinctions may be experiences at the lower latitudes.

Changes in surface water availability and run-off will influence the recharging of groundwater supplies and, in the longer term, aquifers. Water quality may also respond to changes in the amount and timing of precipitation. Rising seas could invade coastal freshwater supplies.

Coastal aquifers may be damaged by saline intrusion as salty groundwater rises. Reduced water supplies would place additional stress on people, agriculture, and the environment. Regional water supplies, particularly in developing countries, will come under many stresses in the 21st century. Global warming will exacerbate the stresses caused by pollution and by growing populations and economies. The most vulnerable regions are arid and semi- arid areas, some low-lying coasts, deltas, and small islands.

Water availability is an essential component of human welfare and productivity. Much of the world's agriculture, hydroelectric power production, water needs and water pollution control is dependent upon the hydrological cycle, and the natural recharching of surface and groundwater resources.

Changes in the natural water availability as a result of global warming would result in impacts which are generally most detrimental in regions already under existing climatic stresses. Even in more benign climates, the effective management of water resources will receive increasing attention as climate change increases the level of competition between potential users for water.

THE TRUTH ABOUT GLOBAL WARMING

On both sides of this multi-faceted global warming debate, data is often misrepresented or misused, scientific facts and processes are contorted, and conclusions from both supporters and opponents are hotly contested. While the politicians and mainstream media generally provide a case to support man made global warming, it is the job of individuals with no political or financial bias to sort through the fact and fiction. In an age where media propaganda is rife, that may be no easy task.

To begin with, climate is not just the 'weather' conditions of a particular town or city, it is a pattern closely calculated within a period of every 30 years.

It is comparative record and analysis of precipitation, radiations, temperatures, winds, and other meteorological conditions. In the 21st century, our climatic conditions are undergoing a sea change.

The term 'global warming' has been an often-discussed term in the UN councils and world environment meetings. Both nature and humans have been equally responsible in bringing about the climatic changes. Though we cannot blame the nature as it has to take its course invariably, but we can surely do our bit in saving the planet.

Let us begin by analyzing how nature and man-made causes were responsible for global warming. Natural Causes: It has been said that nature completes its cycle every 40,000 years and in between, we have undergone

Effects of Ozone Depletion and Global Warming

Ice Age and evolution of humankind to its present form. What changes have occurred that is destroying our planet right now?

- Continent Shifts- 200 million years ago, the 7 continents was a large piece of landmass known as 'Pangaea'. When the continents broke up, it affected the physical features of the landmass, the climate, and oceanic currents. Till today, the Himalayan range constantly grows and is heading towards Asian landmass.
- Volcanic Eruption: When volcanoes erupt, they leave behind a large trail of gases, dust, vapor, and ash. The elements remain in the stratosphere for innumerable years. These create a cloud in the atmosphere and block the sunrays. When the solar radiations do not reach the atmosphere, cooling sets in. This situation occurred in the year 1816 when the Tambora volcano erupted in 1815. The cooling effects lead to frosty summers over Western Europe, Canada and US.
- Oceanic Currents: 71% of the earth consists of water. The oceans soak up most of the sunlight reaching earth. The ocean currents do not remain static; they govern the heat and cold balance throughout the Earth. The flow of the currents determines the coldness of countries like North Atlantic and Alaska. So, when the heat absorbed by water rise up as water vapor (a greenhouse gas), it leads to cloud formation and gives a cooling effect. This phenomenon was exactly what took place during the last Ice Age (14,000 years ago).

Man-made Causes: Difficulty lies in actually pointing the exact cause when man became the destroyer of nature, and indirectly of climate. Even a 0.1 per cent of change in natural patterns disrupts the ecological system. Man has been using nature for his benefits since the times we leant agriculture. One definite period of drastic climatic changes would be the beginning of Industrial Revolution. Much as it had benefited us in terms of technological advancement, it robbed us of nature and we are still paying for its consequences. How does man contribute towards climatic changes?

THE ANSWER IS SIMPLE- GLOBAL WARMING

Global Warming is the "increase in the average temperature of the Earth's near-surface air and oceans since the mid-20th century and its projected continuation". Greenhouse gases, like carbon dioxide, contribute to global warming heavily. Man contributes to global warming in the form of deforestation, pollution, unchecked population growth, and by harming the ecological system. These are explained briefly below:

- *Deforestation:* Trees, or greenery, in a much broader term, are a prerequisite condition to protect the earth. Greenery covers the earth and maintains the balance in atmosphere through the cycle of inhaling carbon dioxide and exhaling oxygen for us to breathe. Felling of tree has been prevalent since time immemorial but with time, man has become reckless. Deforestation disrupts the ecological cycle of the natural world. Without the green cover, how is the carbon dioxide going to process itself?

- *Pollution:* Deforestation meant more land, and with technological advancements, more land meant urbanization of cities. Remember- industrial revolution? With factory productions and uninhibited burning of fossil fuels, man only contributed woe to nature. Climate began to change. Of course, climate changes are not instant, it took and 100s of years and an equal amount of time for man to understand the repercussions of his handiwork. Carbon emissions continue today through the medium of cars, burning electricity, and others.
- *Population:* Two things strike here- transportation and food production. Food production means increase in the level of methane. Why? It is because animals are put into use for producing food. A large population means increase in production of cars and therefore, more burning of fossil fuels like petroleum.
- *Ecological System:* Do you have any explanation as to why daily 100s of species are on the verge of extinction? Humans and animals alike have particular ecological systems founded upon million of years of evolutionary processes. However, through deforestation, decrease in forest lands, basically, the lack of green cover, we have been ravishing the habitat of the animal kingdom in the name of civilization. Similarly, excessive fishing is making them extinct. Deposit of factory spoils on the sea or ocean bed has been poisoning the water habitat. It will not be a surprise if within few years, the plant kingdom and the water bodies die out permanently.

All the factors explained above are contributing to the increase of global warming. You will notice that some areas are getting warmer and some are equally getting colder, which means the seasonal balance have been tilting. The marauding effects of humankind on nature would take another hundreds of years to set it back on track, which would be just the beginning. In addition, we have to begin at the earliest. We need to give nature back what we have taken away from her unflinchingly.

CONTROL THE HARM ALREADY DONE

First, nature cannot be controlled. What we can do now is to set in motion the reverse steps that can rectify the harm done to climatic conditions.

Nations and countries should be standing together to restore the much cherished and desired 'greenery' all over the earth. The main problem as to why we least worried about nature and its climatic changes is because of the lackadaisical, callous, and least bothered attitude of our own.

There is a lack of motivation to protect our environment, our lovable planet. Undeniably, man has been putting off taking note of the depleting green cover until the last moment. Deducting quite a handful of volunteers and government working organizations, who has seriously though upon this matter? WE think that saving nature and reversing the climate is not our prerogative. Instead, we leave it upon governmental organizations and we blame them for any wrongdoings. How tragic could it get? The keyword here is global cooperation.

There are some basic things man can take course too, which will help in restoring the lost glory of our precious environment. So, let's see what we can do to benefit nature?

- Begin to recycle home waste. Separate the renewable and non-renewable wastes. Use the organic material as compost for plants. Use electricity when required only. More than saving you few dollars, it will reduce environmental pollution. Grow plants and shrubs at your home to contribute to the green cover.
- Take to cycles or battery mode bikes as a way of transportation. This will not only reduce carbon emissions, and thereby pollution, it will clear the environment, which will in effect have an impact on the climate. A cleaner environment means a safer atmosphere. If you cannot do without cars, then use alternate sources of energy.
- It is necessary to preserve the fossil fuels. It had taken million of years to form the fossils, which we are using now and so, it will take another few millions to form fossil fuels. In addition, reduction in carbon emission will balance the increasing world average temperature. Reduce dependence upon non-biodegradable sources of energy like plastics.
- Imbalanced climate affects the health too. Respiratory problems, waterborne diseases, mosquito- borne diseases would soon become a common phenomenon if not taken care of. Try to maintain a pure and clean environment at home.
- Encourage your company to contribute in carbon footprint assessment. Only an assessment would help to devise strategies to overcome increase carbon exposure.

These steps and many more, actually innumerable ways, are there to reverse the climatic degradation only if man becomes more aware and understands the repercussions of environmental hazards seriously. The climate reversal techniques will not only reduce the increasing global warming but also reduce the temperature, thus balancing the heat with the cold. Overall, the precautionary measures upheld now will lay a precedent for the future generations to take care of the greenery of nature and its bounty. Humans, plant kingdom, and animal kingdom are an organic species. We need to learn to share the ecological system with other life forms.

Global warming is (presently) demonstrated in increases in global average air and ocean temperatures and widespread melting of snow and ice. This is most likely caused by the observed increase in anthropogenic greenhouse gas concentrations in the atmosphere. Global warming will almost certainly continue due to the time scales associated with climate processes, even if greenhouse gas concentrations were to be stabilised at present levels. The increasing temperatures and the melting of snow and ice are leading to rising global average sea levels. The uptake of anthropogenic carbon is leading to increasing acidification of the oceans. The frequencies and intensities of extreme weather are changing.

Weather extremes – such as hot and cold temperatures, storms and cyclones, and heavy rainfalls with risks of flooding will increase in frequency in many regions. Seasons are changing, with early spring and late autumn. Plant and animal ranges are likely to be shifting and some plant and animal populations are likely to decline. A warmer climate is favourable for the survival of pathogenic microorganisms and parasites, and vector habitats are expanding geographically, leading to new human and animal disease scenarios. Even today, we see impacts of climate change in many natural and man made systems, and these impacts will probably become more and more visible in the coming decades. Examples of systems and sectors most likely to be affected are ecosystems, food, fibre and forest products, coastal systems and low-lying areas, polar and boreal areas, settlements and industries linked to climate-sensitive resources, handling of water (drinking, waste, urban runoff) in cities. Human and animal health aspects are also critical.

Climate change will almost certainly affect many vital functions in society as well as in the natural environment, and will pose a major threat to global sustainable development. Responses to climate change are therefore imperative. This entails a continuous risk management process, where combinations of adaptation and mitigation measures can reduce the risks associated with climate change. Research funded by Formas shall contribute to an increased capacity in society and ecosystems to respond to climate change.

Many questions regarding the functions and interactions of various systems related to climate change are still to be answered in order to improve our capacity to respond to climate change.

Research is needed on the interactions between natural systems and man made infrastructure, for instance in urban planning and construction. Moreover, research is essential in the development of effective measures for mitigation and adaptation. Analyses of interactions between natural and human systems are necessary, as well as analyses of interactions between different societal interests and processes. Analyses of equity and burden sharing between people and countries, generations, and sectors are also needed.

Chapter 5

Effects of Global Warming on Ecosystems

Although not every scientist worldwide may look at global warming in the same way, they do overwhelmingly agree that the Earth's atmosphere is getting warmer. Worldwide temperatures have risen more than 1°F over the past century, and 17 of the past 20 years have been the hottest ever recorded. A special report issued by Time magazine on April 3, 2006, the Intergovernmental Panel on Climate Change in their third report, released in 2001, had analysed data from the past two decades representing properties such as air and ocean temperatures and the habitat characteristics and patterns of wildlife. Examples of observed changes included"shrinkage of glaciers, thawing of permafrost, later freezing and earlier breakup of ice on rivers and lakes, lengthening of mid- to high-latitude growing seasons, poleward and altitudinal shifts of plant and animal ranges, declines of some plant and animal populations, and earlier flowering of trees, emergence of insects, and egg- laying in birds. Associations between changes in regional temperatures and observed changes in physical and biological systems have been documented in many aquatic, terrestrial, and marine environments."

The IPCC is an organized group of more than 2,500 climate experts from around the world that consolidates their most recent scientific findings every five to seven years into a single report, which is then presented to the world's political leaders. The IPCC was established in 1988 by the World Meteorological Organization and the United Nations Environment Programme to specifically address the issue of global warming. As a result of their comprehensive analysis, they have determined that this steady warming has had a significant impact on at least 420 animal and plant species and also on natural processes. Furthermore, this has not just occurred in one geographical location but worldwide. In the IPCC's fourth report, released in February 2007, they concluded that it is"very likely" that heat-trapping emissions from human

Effects of Global Warming on Ecosystems

activities have caused"most of the observed increase in globally averaged temperatures since the mid-20th century."

Also, in the February 2007 report, they concluded the following:

- "Human induced warming over recent decades is already affecting many physical and biological processes on every continent. Nearly 90 per cent of the 29,000 observational data series examined revealed changes consistent with the expected response to global warming, and the observed physical and biological responses have been the greatest in the regions that have warmed the most."

In these studies, scientists have been able to break down the natural and human-caused components in order to see how much of an effect humans have had. Human effects can include activities such as burning fossil fuels, agricultural practices, deforestation, industrial processes, the introduction of invasive plant or animal species, and various types of land-use change. In many cases, scientists do not need to look very far to see the effects a warming world is having on the environment and the Earth's ecosystems. Glaciers worldwide are melting at an accelerated rate never seen before. The cap of ice on top of Kilimanjaro is rapidly disappearing, the glaciers of world-renowned Glacier National Park in the United States and Canada are melting and projected to be gone in the next few decades, and the glaciers in the European Alps are experiencing a similar fate. In the world's tropical oceans, vast expanses of beautiful, brilliantly coloured coral reefs are dying off as oceans slowly become too warm.

Unable to survive the higher temperatures, the corals are undergoing a process called bleaching and are turning white and dying. In the Arctic, as temperatures climb, ice is melting at accelerated rates, leaving polar bears stranded, destroying their feeding and breeding grounds, and causing them to starve and drown. Permafrost is melting at accelerated rates. As the ground thaws, it is disrupting the physical and chemical components of the ecosystem by causing the ground to shift and settle, toppling buildings and twisting roads and railroad tracks, as well as releasing methane gas into the atmosphere. Weather patterns are also changing. El Niño events are triggering destructive weather in the eastern Pacific. There has been an increase in extreme weather events, such as hurricanes. Droughts have become more prevalent in some geographical areas, such as parts of Asia, Africa, Australia, and the American Southwest.

Animal and plant habitats have been disrupted, and, as temperatures continue to climb, there have been several documented migrations of individual species moving northward or to higher elevations on individual mountain ranges. Migration patterns are also being affected, such as those already documented of beluga whales, butterflies, and polar bears. Spring is also arriving earlier in some areas, which is now influencing the timing of bird and fish migration, egg laying, leaf unfolding, and spring planting for agriculture. In fact, based on satellite imagery documentation of the Northern Hemisphere, growing seasons have steadily become longer since 1980. While species have been faced with changing environments in the past and have been able to adapt in many cases, the IPCC climate change scientists view this current rate of change with alarm. They fully expect the magnitude of these changes to increase with the temperatures over the next century and beyond.

The concern is that many species and ecosystems will not be able to adapt as rapidly as the effects of global warming will cause the environment to change. In addition, there will also be other disturbances, such as floods, insect infestations, and the spread of disease, wildfire, and drought. Any of these additional challenges can destroy a species or habitat. In particular, alpine and polar species are especially vulnerable to the effects of climate change because as species move northward or higher on mountains, these species' habitats will shrink, leaving them with nowhere to go. With so much evidence, most scientists no longer doubt that global warming is real, nor do they question the fact that humans are to blame.

All it takes is a look at the air quality over significant population and industrial centres to begin to grasp the effect that humans can have on the environment. Based on temperature records kept before the beginning of the Industrial Revolution, carbon dioxide in the atmosphere has increased 30 per cent above those earlier levels. Not only are the levels higher, but they increase annually. The IPCC, at current conditions, by the year 2100, the average temperature is expected to increase between $2°$ and $11.5°F$- an amount more than 50 per cent higher than what was predicted only 50 years ago. Within the IPCC's predicted temperature range, at the lower end, storms would become more frequent and intense, droughts would be more severe, and coastal areas would be flooded by rising sea levels from melting glaciers and ice caps. There would be enough of a disruption that ecosystems worldwide would be thrown out of balance and altered. If, however, the temperature rise falls towards the higher end of the estimate, the results on ecosystems worldwide would be disastrous.

Sea levels could rise so much that entire islands of low elevation, such as the Maldives, could completely disappear. Other areas, such as the Nile Delta and much of the United States' coastal southeast could become completely uninhabitable. Climate zones could shift, completely disrupting land-use practices. For instance, the current agricultural region of the Great Plains in the United States could be shifted to Canada. The southern portion of the United States could become more like Central or South America. Siberia would no longer be a frozen, desolate landscape. Parts of Africa could become dry, desolate wastelands. If this were to happen, it would have a severe impact on the production of agriculture.

Areas currently equipped to produce agriculture would no longer be able to, and areas that were able, based on climate, may not have the financial resources or the proper soils. The ripple effect of these disruptions would be felt worldwide. Millions of people would be forced to migrate from newly uninhabitable regions to new areas where they could survive. This would also affect public health. Rising seas would contaminate freshwater with salt water; there would be more heat-related illnesses and deaths; and disease-carrying rodents and insects, such as mice, rats, mosquitoes, and ticks, would spread diseases such as malaria, encephalitis, Lyme disease, and dengue fever. Scientists of the IPCC agree that one of the most serious aspects of all this drastic change is that it is happening so fast. These changes are happening at a faster pace than the Earth has seen in the last 100 million years. While humans may be able to pick up and move to a new location, animals and their associated ecosystems cannot. The choices people make and the actions they take today will determine the fate of other life and their ecosystems tomorrow.

RESULTS OF GLOBAL WARMING ON ECOSYSTEMS

In a study on the effects of global warming on the Earth's ecosystems conducted by Chris Thomas, a conservation biologist at the University of Leeds in the United Kingdom, he states "Climate change now represents at least as great a threat to the number of species surviving on Earth as habitat destruction and modification." Thomas worked with a group of 18 scientists worldwide in the largest study of its type ever accomplished. The end result of their study came down to this conclusion: "By 2050, rising temperatures, made more severe through human-induced input, such as the burning of fossil fuels, could send more than a million of Earth's land-dwelling plants and animals down the road to extinction." The research team worked by themselves in six biodiversity-rich areas around the world, ranging from Australia to South Africa. As they gathered field data about species distribution and regional climate, they programmed the information into computer climate models.

The purpose of the computer models was to simulate the direction and distance individual species would migrate in response to temperature and climate changes. Once all team members had collected their specific data, they combined it into a single model in order to understand the global concept.

Once the model had been carefully evaluated, it was determined that by the year 2050, at predicted global warming rates, 15 to 37 per cent of the 1,103 species studied could be at risk of extinction. When the study area was expanded to cover the entire Earth, the researchers estimated that worldwide more than a million species could begin to face extinction by 2050."This study makes clear that climate change is the biggest new extinction threat," said coauthor Lee Hannah of Conservation International in Washington, D.C."In some cases we found that there will no longer be anywhere climatically suitable for these species to live; in other cases they may be unable to reach distant regions where the climate will be suitable.

Other species are expected to survive in much reduced areas, where they may then be at risk from other threats," said coauthor Guy Midgley of the National Botanical Institute in Cape Town, South Africa."Seeing the range of responses across all 1,103 species, it becomes obvious that we have a lot of work left to do before we can accurately predict what types of animals and plants are most at risk. This range of responses shows that species will not be able to move as whole biological communities, and that the typical natural communities we recognize today will probably not exist under future conditions. Figuring out what will replace them requires a lot of imagination," said coauthor Alison Cameron of the University of Leeds. Chris Thomas, almost all future climate projections expect more warming and even more extinction between 2050 and 2100, and, even though projections are only made to 2100, temperatures will still keep going up and more warming will occur after that.

This group of researchers say that taking action now to slow global warming is important to make sure that climate change ends up on the low end of the prediction in order to avoid"catastrophic extinctions." Thomas also stated that because there may be a time lag between the climate changing and the last individual of a species dying off, the rapid reduction of greenhouse gas emissions may enable some species to survive. It is also important to keep in mind that although some species may be able to migrate successfully to a new location, some plant and animal species that live in high mountain or polar ecosystems cannot move farther to escape warming temperatures. Similarly, coral reef systems cannot just pick up and move to a new location.

Long-established communities have remained where they are in order to survive. Robert Puschendorf, a biologist at the University of Costa Rica, believes these estimates"might be optimistic." As global warming interacts with other factors such as habitat destruction, increase in invasive species, and the buildup of carbon dioxide in the environment, there may be more than a million species that face extinction. In connection with Chris Thomas's study, Richard J. Ladle, *et al.*, from Oxford University"Dangers of Crying

Wolf Over Risk of Extinctions," bringing up an issue important to all scientific investigations. It cautioned that the scientific world needs to"learn how to deal with increasingly sensationalist mass media."

They warn that policy-makers must be informed by a"balanced assessment of scientific knowledge and not popular perception created by commercially driven media. Departure from rational objectivity risks undermining public trust in the natural sciences and could play into the hands of antienvironmentalists. This places responsibilities on both scientists and journalists to ensure fair and accurate reporting of their work." Ladle's report clarified that a large portion of the media incorrectly focused on the idea that more than a million species would definitely go extinct by 2050 when the study actually concluded that the extinctions will occur eventually and not in the next 50 years. They reported that 21 out of 29 substances quoting Thomas's study misinformed the public by inferring the species would be extinct by 2050. Their chief concern is that the media, or anyone else, through unclear presentation, could inadvertently increase public cynicism and complacency about climate change and biodiversity loss. The resulting message is that when valuable studies, such as Chris Thomas's, are conducted and the results are released, in order to receive the best results from the scientific community, policy-makers, and the public in general, it is critical not to sensationalize or misrepresent the facts. Otherwise, it puts the real message in jeopardy and may even do more harm than good. Both plant and animal species are at risk due to the effects of global warming.

The World Wildlife Fund, the golden toad and the harlequin frog of Costa Rica have already disappeared as a direct result of global warming. As different components of an ecosystem change, it can upset the natural balance in many ways. For instance, it can disrupt a species by having spring show up a week or two earlier. Over time, delicate balances have been set up between animals and the food they eat. If a certain animal relies on a specific food, but global warming has already caused the food to grow through its life cycle before the animal is ready to use it, then it will have a direct negative effect on that animal-the food will not be available when the animal needs it. That animal's health and existence may be threatened. Furthermore, this could cause a ripple effect in the food chain, having an impact on more than one species.

One example of this is when spring comes earlier than it has in the past. The timing of feeding for newly hatched birds may not correspond to the availability of worms or insects, impacting the fledglings' chances of survival. The WWF, climate records compared with long-term records of flowering and nesting times show a noticeable shift away from each other, depicting global warming trends. In Britain, flowering time and leaf-on records, which date back to 1736, have provided concrete evidence of climate-related seasonal shifts. Long-term trends towards earlier bird breeding, earlier spring migrant arrival, and later autumn departure dates have also been recorded in North America. Changes in migratory patterns have also been documented in Europe.

Once again, the WWF documents that climate change and global warming are currently affecting species in many ways. Animals and plants that require cooler temperatures in order to survive, which need to either migrate northward or higher in elevation in a mountain ecosystem, are already being documented in several places worldwide. This is occurring in the European Alps, in Queensland in Australia, and in the rain forests of Costa Rica. Fish in the North Sea have been documented migrating northward. Fish populations that used to inhabit areas around Cornwall, England, have migrated as far north as the Shetland and Orkney Islands. WWF global warming experts believe, based on this evidence, that "the impacts on species are becoming so significant that their movements can be used as an indicator of a warming world.

They are silent witnesses of the rapid changes being inflicted on the Earth." In fact, these same scientists believe that global warming could begin causing extinctions of animal species in the near future because the heating caused by accelerated global warming has a severe impact on the Earth's many delicate ecosystems-both on the land and the species that live on it. Worldwide, there are species and habitats that have now been identified as being threatened and endangered due to the effects of global warming. Because ecosystems can be altered to the point where the damage becomes irreversible and species must either adapt to survive or face extinction, it is critical that the issue of global warming be addressed and acted upon now before it is too late.

OBSERVED AND EXPECTED EFFECTS ON ECOSYSTEMS

There are several pieces of physical evidence that scientists have already identified indicating that global warming is already in progress and affecting all the ecosystems on Earth. By monitoring the health of ecological conditions, scientists can see the effects climate change is having on the individual components that comprise the ecosystem. Because an ecosystem is such a tightly knit system of living things within their natural environment, if one component is affected, a ripple effect can be started, eventually endangering the entire ecosystem.

POLAR AND ICE-RELATED CHANGES

Of all the Earth's ecosystems, climate change in the polar regions is expected to be more rapid and more severe than anywhere else. If snow and ice are melted, this will greatly change the albedo of the environment.

As darker surfaces increase, more sunlight will be absorbed, rapidly heating up the Earth's surface and atmosphere. In addition, as worldwide species continue to migrate northward under warming temperatures, thick dark vegetation will crowd areas that were once wideopen snowfields, also lowering the albedo. The IPCC, the average annual temperature in the Arctic has increased by 1.67°F over the past century—which equals a rate roughly twice as fast as the global average. Winter temperatures have been consistently 3.3°F warmer over the past century. The effects of this warming have been seen in decreases in thickness and extent of sea ice, the melting of permafrost, and later freezing and earlier breakup dates of winter sea ice.

Glaciers worldwide are also melting: There are glaciers on all the Earth's continents except Australia and at all latitudes from the Tropics to the polar regions. There is widespread evidence that glaciers are retreating in many areas of the world. Because sea ice in the polar regions is breaking up earlier in the year, polar bears and walrus are already suffering. Their feeding and breeding grounds are disappearing, their territorial boundaries are gone, and they are dying. Early breakup is also affecting the hunting habits and lifestyles of northern native inhabitants, forcing many to abandon lifelong rich, traditional cultures and relocate to other areas.

FIRE AND DROUGHT

Over large areas of the Earth, nights have warmed up more than days have. In fact, since 1950, minimum temperatures on land have rapidly increased. An increase in warm temperatures will lead to increases in the number of heat waves that strike urban areas, which will cause more heat-related illnesses and deaths. Global warming is causing a more intense hydrologic cycle with increased evaporation. The greater the evaporation rates, the more soils and vegetation will dry out. As temperatures rise and vegetation dries out, areas will become drier under droughtlike conditions, and wildfires will become more common. This occurred during the 2007 summer and fall in California. During this tragic event, more than 772 square miles of land burned from Santa Barbara County to the U.S.-Mexico border. Nearly 100 people were injured, 9 died, and more than 1,500 homes were destroyed. Wildfires forced 265,000 residents to flee their homes.

BIOLOGICAL CHANGES

Global warming also affects the occurrence and spread of disease. It makes large populations vulnerable if the pathogen is spread quickly. Warmer temperatures and more precipitation will help spread disease organisms from rodents and insects to larger areas. The world's undeveloped countries are expected to be hit the hardest. Rising sea levels due to the melting of glaciers and ice caps are causing coral to lose the symbiotic algae that they must have for nutrition. This algae is also what gives them their beautiful, vibrant colours. When the algae die, the coral looks white and is referred to as bleached.

It only takes a small increase in temperature above normal summer levels for periods of time as short as only two or three days to cause this reaction. In 1998, one of the hottest years on record, coral reefs worldwide experienced the most extensive bleaching ever recorded. Coral bleaching was reported in 60 countries and island nations in the Indian Ocean, Red Sea, Pacific Ocean, Persian Gulf, Mediterranean, and Caribbean. Plant and animal species are located where the climatic factors enable them to thrive. If any of these characteristics change-as they do with global warming-then species will attempt to migrate. Whether they are successful or not depends on several factors: the rate of change, availability of acceptable habitat, a physical way to relocate to acceptable habitat, and avoidance of predators.

If any of these factors works out wrong, the species can become threatened, endangered, or extinct. NASA, based on data obtained through satellite observation over the past 20 years, areas in both North America and Eurasia have longer growing seasons associated with the buildup of greenhouse gases in the atmosphere. During the course of data collection, NASA scientists noticed dramatic changes in the timing of when leaves first appeared in the spring and when they fell off in the fall. By monitoring when things turn green, NASA scientists determined that in Eurasia the growing season is currently almost 18 days longer than it was two decades ago. Today, spring arrives a week earlier and autumn 10 days later than they did in the past.

PHYSICAL CHANGES

The rise in mean global surface temperature has caused spring to come earlier in many parts of the world. This has led to a longer growing season in middle and high-latitude areas. The effects of this are widespread- leaves come out earlier and stay longer, and breeding and migration patterns of wildlife are affected. For instance, in the northeastern United States, the frost-free season now begins approximately 11 days earlier than it did in the 1950s. As worldwide temperatures climb, the hydrologic cycle will intensify, producing more intense phenomenon such as flooding, landslides, and erosion. The areas at highest risk are those at mid to high latitudes. Trends are already apparent in several polar locations in the Northern Hemisphere. P. Y. Groisman and D. R. Easterling of the NCDC, over the past few decades snowfall has increased about 20 per cent over northern Canada and about 11 per cent over Alaska. An increase in snowfall has also been observed over China. T. R. Karl and R. W. Knight, observations for the last century indicate that extreme weather events in the United States have increased by about 20 per cent. Increases in heavy rainfall have also been reported in Japan and northeastern Australia.

Tidal gauges are placed worldwide and mean sea level is monitored. Over the past century, global mean sea level has risen 4 to 10 inches with the average being 7 inches. This rate is greater than what the average has been over the past few thousand years. The IPCC, although unsure how much ocean levels will rise in the next century, has projected that the rate will be at least two to four times the rate of the last century. The reason for the uncertainty is that the behaviour of the Antarctic and Greenland ice sheets remains uncertain at this point. If their melting rate increases, future sea-level rise will most likely be on the larger side of the projection. Sea-level rise and coastal flooding are also governed by wind and pressure patterns, ocean circulation, and the characteristics of the coastline-whether there are coastal wetlands, beaches, islands, or other structures to act as barriers.

ECOSYSTEM SERVICES IN GLOBAL WARMING

Humankind benefits from a multitude of resources and processes that are supplied by natural ecosystems. Collectively, these benefits are known as ecosystem services and include products like clean drinking water and processes such as the decomposition of wastes. While scientists and environmentalists have discussed ecosystem services for decades, these services were popularized and their definitions formalized by the United Nations 2004 Millennium Ecosystem Assessment, a four-year study involving four broad categories: provisioning, such as the production of food and water; regulating, such as the control of climate and disease; supporting, such as nutrient cycles and crop pollination; and cultural, such as spiritual and recreational benefits.

As human populations grow, so do the resource demands imposed on ecosystems and the impacts of our global footprint. Natural resources are not invulnerable and infinitely available. The environmental impacts of anthropogenic actions, which are processes or materials derived from human activities, are becoming more apparent – air and water quality are increasingly compromised, oceans are being overfished, pests and diseases are extending beyond their historical boundaries, and deforestation is exacerbating flooding downstream.

It has been reported that approximately 40-50 per cent of Earth's ice-free land surface has been heavily transformed or degraded by anthropogenic activities, 66 per cent of marine fisheries are either overexploited or at their limit, atmospheric CO_2 has increased more than 30 per cent since the advent of industrialization, and nearly 25 per cent of Earth's bird species have gone extinct in the last two thousand years. Society is increasingly becoming aware that ecosystem services are not only limited, but also that they are threatened by human activities. The need to better consider long-term ecosystem health and its role in enabling human habitation and economic activity is urgent. To help inform decision-makers, many ecosystem services are being assigned economic values, often based on the cost of replacement with anthropogenic alternatives. The ongoing challenge of prescribing economic value to nature, for example through biodiversity banking, is prompting transdisciplinary shifts in how we recognize and manage the environment, social responsibility, business opportunities, and our future as a species.

The simple notion of human dependence on Earth's ecosystems probably reaches to the start of our species' existence, when we benefited from the products of nature to nourish our bodies and for shelter from harsh climates. Recognition of how ecosystems could provide more complex services to mankind date back to at least Plato who understood that deforestation could lead to soil erosion and the drying of springs.

However, modern ideas of ecosystem services probably began with Marsh in 1864 when he challenged the idea that Earth's natural resources are not infinite by pointing out changes in soil fertility in the Mediterranean. Unfortunately, his observations and cautions passed largely unnoticed at the time and it was not until the late 1940s that society's attention was again brought to the matter. During this era, three key authors – Osborn, Vogt, and Leopold – awakened and promoted recognition of human dependence on the environment with the idea of 'natural capital'.

In 1956, Sears drew attention to the critical role of the ecosystem in processing wastes and recycling nutrients. An environmental science textbook called attention to "the most subtle and dangerous threat to man's existence... the potential destruction, by man's own activities, of those ecological systems upon which the very existence of the human species depends".

The term 'environmental services' was finally introduced in a report of the Study of Critical Environmental Problems, which listed services including insect pollination, fisheries, climate regulation and flood control. In following years, variations of the term were used, but eventually 'ecosystem services' became the standard in scientific literature.

Modern expansions of the ecosystem services concept include socio-economic and conservation objectives, which are discussed below. For a more complete history of the concepts and terminology of ecosystem services.

ECOLOGY

Understanding of ecosystem services requires a strong foundation in ecology, which describes the underlying principles and interactions of organisms and the environment. Since the scales at which these entities interact can vary from microbes to landscapes, milliseconds to millions of years, one of the greatest remaining challenges is the descriptive characterization of energy and material flow between them.

For example, the area of a forest floor, the detritus upon it, the microorganisms in the soil and characteristics of the soil itself will all contribute to the abilities of that forest for providing ecosystem services like carbon sequestration, water purification, and erosion prevention to other areas within the watershed. Note that it is often possible for multiple services to be bundled together and when benefits of targeted objectives are secured, there may also be ancillary benefits – the same forest may provide habitat for other organisms as well as human recreation, which are also ecosystem services.

The complexity of Earth's ecosystems poses a challenge for scientists as they try to understand how relationships are interwoven among organisms, processes and their surroundings. As it relates to human ecology, a suggested research agenda for the study of ecosystem services includes the following steps:

1. Identification of ecosystem service providers – species or populations that provide specific ecosystem services – and characterization of their functional roles and relationships;
2. Determination of community structure aspects that influence how ESPs function in their natural landscape, such as compensatory responses that stabilize function and non-random extinction sequences which can erode it; of services;
3. Measurement of the spatial and temporal scales ESPs and their services operate on.

Recently, a technique has been developed to improve and standardize the evaluation of ESP functionality by quantifying the relative importance of different species in terms of their efficiency and abundance. Such parametres provide indications of how species respond to changes in the environment and are useful for identifying species that are disproportionately important at providing ecosystem services.

However, a critical drawback is that the technique does not account for the effects of interactions, which are often both complex and fundamental in maintaining an ecosystem and can involve species that are not readily detected as a priority.

Even so, estimating the functional structure of an ecosystem and combining it with information about individual species traits can help us understand the resilience of an ecosystem amidst environmental change.

Many ecologists also believe that the provision of ecosystem services can be stabilized with biodiversity. Increasing biodiversity also benefits the variety of ecosystem services available to society. Understanding the relationship between biodiversity and an ecosystem's stability is essential to the management of natural resources and their services.

The Redundancy Hypothesis

The concept of ecological redundancy is sometimes referred to as functional compensation and assumes that more than one species performs a given role within an ecosystem. More specifically, it is characterized by a particular species increasing its efficiency at providing a service when conditions are stressed in order to maintain aggregate stability in the ecosystem.

However, such increased dependence on a compensating species places additional stress on the ecosystem and often enhances its susceptibility to subsequent disturbance. The redundancy hypothesis can be summarized as "species redundancy enhances ecosystem resilience".

The Rivet Hypothesis

Another idea uses the analogy of rivets in an airplane wing to compare the exponential effect the loss of each species will have on the function of an ecosystem; this is sometimes referred to as rivet popping. If only one species disappears, the loss of the ecosystem's efficiency as a whole is relatively small; however if several species are lost, the system essentially collapses as an airplane wing would, were it to lose too many rivets.

The hypothesis assumes that species are relatively specialized in their roles and that their ability to compensate for one another is less than in the redundancy hypothesis. As a result, the loss of any species is critical to the performance of the ecosystem. The key difference is the rate at which the loss of species affects total ecosystem function.

The Portfolio Effect

A third explanation, known as the portfolio effect, compares biodiversity to stock holdings, where diversification minimizes the volatility of the investment, or in this case, the risk in stability of ecosystem services. This is related to the idea of response diversity where a suite of species will exhibit differential responses to a given environmental perturbation and therefore when considered together, they create a stabilizing function that preserves the integrity of a service.

Several experiments have tested these hypotheses in both the field and the lab. In ECOTRON, a laboratory in the UK where many of the biotic and abiotic factors of nature can be simulated, studies have focused on the effects of earthworms and symbiotic bacteria on plant roots. These laboratory experiments seem to favour the rivet hypothesis. However, a study on grasslands at Cedar Creek Reserve in Minnesota seems to support the redundancy hypothesis, as have many other field studies.

ECONOMICS

There is extensive disagreement regarding the environmental and economic values of ecosystem services. Some people may be unaware of the environment in general and humanity's interrelatedness with the natural environment, which may cause misconceptions. Although environmental awareness is rapidly improving in our contemporary world, ecosystem capital and its flow are still poorly understood, threats continue to impose, and we suffer from the so-called 'tragedy of the commons'.

Many efforts to inform decision-makers of current versus future costs and benefits now involve organizing and translating scientific knowledge to economics, which articulate the consequences of our choices in comparable units of impact on human well-being. An especially challenging aspect of this process is that interpreting ecological information collected from one spatial- temporal scale does not necessarily mean it can be applied at another; understanding the dynamics of ecological processes relative to ecosystem services is essential in aiding economic decisions.

Weighting factors such as a service's irreplaceability or bundled services can also allocate economic value such that goal attainment becomes more efficient.The economic valuation of ecosystem services also involves social communication and information, areas that remain particularly challenging and are the focus of many researchers. In general, the idea is that although individuals make decisions for any variety of reasons, trends reveal the aggregative preferences of a society, from which the economic value of services can be inferred and assigned.

The six major methods for valuing ecosystem services in monetary terms are:
1. Avoided cost: Services allow society to avoid costs that would have been incurred in the absence of those services (e.g. waste treatment by wetland habitats avoids health costs)
2. Replacement cost: Services could be replaced with man-made systems (e.g. restoration of the Catskill Watershed cost less than the construction of a water purification plant)
3. Factor income: Services provide for the enhancement of incomes (e.g. improved water quality increases the commercial take of a fishery and improves the income of fishers)
4. Travel cost: Service demand may require travel, whose costs can reflect the implied value of the service (e.g. value of ecotourism experience is at least what a visitor is willing to pay to get there)
5. Hedonic pricing: Service demand may be reflected in the prices people will pay for associated goods (e.g. coastal housing prices exceed that of inland homes)
6. Contingent valuation: Service demand may be elicited by posing hypothetical scenarios that involve some valuation of alternatives (e.g. visitors willing to pay for increased access to national parks)

MANAGEMENT AND POLICY

Utility of Radiative Forcing: The TAR and other assessments have concluded that RF is a useful tool for estimating, to a first order, the relative global climate impacts of differing climate change mechanisms. In particular, RF can be used to estimate the relative equilibrium globally averaged surface temperature change due to different forcing agents. However, RF is not a measure of other aspects of climate change or the role of emissions. Previous GCM studies have indicated that the climate sensitivity parameter was more or less constant between mechanisms.

However, this level of agreement was found not to hold for certain mechanisms such as ozone changes at some altitudes and changes in absorbing aerosol. Because the climate responses, and in particular the equilibrium climate sensitivities, exhibited by GCMs vary by much more than 25 per cent, Ramaswamy and Jacob concluded that RF is the most simple and straightforward measure for the quantitative assessment of climate change mechanisms, especially for the LLGHGs. This part discusses the several studies since the TAR that have examined the relationship between RF and climate response. Note that this assessment is entirely based on climate model simulations.

VERTICAL FORCING PATTERNS AND SURFACE ENERGY BALANCE CHANGES

The vertical structure of a forcing agent is important both for efficacy and for other aspects of climate response, particularly for evaluating regional and vertical patterns of temperature change and also changes in the hydrological cycle. For example, for absorbing aerosol, the surface forcings are arguably a more useful measure of the climate response than the RF. It should be noted that a perturbation to the surface energy budget involves sensible and latent heat fluxes besides solar and longwave irradiance; therefore, it can quantitatively be very different from the RF, which is calculated at the tropopause, and thus is not representative of the energy balance perturbation to the surface-troposphere system.

While the surface forcing adds to the overall description of the total perturbation brought about by an agent, the RF and surface forcing should not be directly compared nor should the surface forcing be considered in isolation for evaluating the climate response. Therefore, surface forcings are presented as an important and useful diagnostic tool that aids understanding of the climate response.

SPATIAL PATTERNS OF RADIATIVE FORCING

Each RF agent has a unique spatial pattern. When combining RF agents it is not just the global mean RF that needs to be considered. For example, even with a net global mean RF of zero, significant regional RFs can be present and these can affect the global mean temperature response. Spatial patterns of RF also affect the pattern of climate response.

However, note that, to first order, very different RF patterns can have similar patterns of surface temperature response and the location of maximum RF is rarely coincident with the location of maximum response. Identification of different patterns of response is particularly important for attributing past climate change to particular mechanisms, and is also important for the prediction of regional patterns of future climate change. This stage employs RF as the method for ranking the effect of a forcing agent on the equilibrium global temperature change, and only this aspect of the forcing-response relationship is discussed.

ALTERNATIVE METHODS OF CALCULATING RADIATIVE FORCING

RFs are increasingly being diagnosed from GCM integrations where the calculations are complex. This stage also discusses several mechanisms that include some response in the troposphere, such as cloud changes. These mechanisms are not initially radiative in nature, but will eventually lead to a radiative perturbation of the surface-troposphere system that could conceivably be measured at the TOA. Jacob refer to these mechanisms as non-radiative forcings. Alternatives to the standard stratospherically adjusted RF definition have been proposed that may help account for these processes.

Since the TAR, several studies have employed GCMs to diagnose the zero- surface-temperature-change RF. These studies have used a number of different methodologies. Shine fixed both land and sea surface temperatures globally and calculated a radiative energy imbalance: this technique is only feasible in GCMs with relatively simple land surface parametrizations. Hansen fixed sea surface temperatures and calculated an RF by adding an extra term to the radiative imbalance that took into account how much the land surface temperatures had responded.

Sokolov diagnosed the zero-surface-temperaturechange RF by computing surface-only and atmospheric-only components of climate feedback separately in a slab model and then modifying the stratospherically adjusted RF by the atmospheric-only feedback component. Gregory used a regression method with a globally averaged temperature change ordinate to diagnose the zero-surface-temperature-change RF: this method had the largest uncertainties. Shine, Hansen and Sokolov all found that that the fixed-surface-temperature RF was a better predictor of the equilibrium global mean surface temperature response than the stratospherically adjusted RF. Further, it was a particularly useful diagnostic for changes in absorbing aerosol where the stratospherically adjusted RF could fail as a predictor of the surface temperature response.

Differences between the zero-surface-temperature-change RF and the stratospherically adjusted RF can be caused by semi-direct and cloud-aerosol interaction effects beyond the cloud albedo RF. For most mechanisms, aside from the case of certain aerosol changes, the difference is likely to be small.

These calculations also remove problems associated with defining the tropopause in the stratospherically adjusted RF definition. However, stratospherically adjusted RF has the advantage that it does not depend on relatively uncertain components of a GCM's response, such as cloud changes. For the LLGHGs, the stratospherically adjusted RF also has the advantage that it is readily calculated in detailed off-line radiation codes. However, to first order, all methods are comparable and all prove useful for understanding climate response.

LINEARITY OF THE FORCING-RESPONSE RELATIONSHIP

Reporting findings from several studies, the TAR concluded that responses to individual RFs could be linearly added to gauge the global mean response, but not necessarily the regional response. Since then, studies with several equilibrium and/or transient integrations of several different GCMs have found no evidence of any non-linearity for changes in greenhouse gases and sulphate aerosol. Two of these studies also examined realistic changes in many other forcing agents without finding evidence of a non-linear response. In all four studies, even the regional changes typically added linearly.

However, Meehl observed that neither precipitation changes nor all regional temperature changes were linearly additive. This linear relationship also breaks down for global mean temperatures when aerosol-cloud interactions beyond the cloud albedo RF are included in GCMs. Studies that include these effects modify clouds in their models, producing an additional radiative imbalance.

Rotstayn and Penner found that if these aerosol-cloud effects are accounted for as additional forcing terms, the inference of linearity can be restored. Studies also find non-linearities for large negative RFs, where static stability changes in the upper troposphere affect the climate feedback.

For the magnitude and range of realistic RFs discussed in this stage, and excluding cloud-aerosol interaction effects, there is high confidence in a linear relationship between global mean RF and global mean surface temperature response.

EFFICACY AND EFFECTIVE RADIATIVE FORCING

Efficacy is defined as the ratio of the climate sensitivity parameter for a given forcing agent to the climate sensitivity parameter for CO2 changes, that Efficacy can then be used to define an effective RF. For the effective RF, the climate sensitivity parameter is independent of the mechanism, so comparing this forcing is equivalent to comparing the equilibrium global mean surface temperature change. That is, $ÄT_s = ëCO_2 \times E_i \times RF_i$ Preliminary studies have found that efficacy values for a number of forcing agents show less model dependency than the climate sensitivity values.

Effective RFs have been used get one step closer to an estimator of the likely surface temperature response than can be achieved by using RF alone. Adopting the zero-surface-temperature-change RF, which has efficacies closer to unity, may be another way of achieving similar goals.

This part assesses the efficacy associated with stratospherically adjusted RF, as this is the definition of RF adopted in this stage. Therefore, cloudaerosol interaction effects beyond the cloud albedo RF are included in the efficacy term. As space is limited not all these studies are explicitly discussed in the main text.

Generic Understanding

Since the TAR, several GCM studies have calculated efficacies and a general understanding is beginning to emerge as to how and why efficacies vary between mechanisms. The initial climate state, and the sign and magnitude of the RF have less importance but can still affect efficacy.

These studies have also developed useful conceptual models to help explain variations in efficacy with forcing mechanism. The efficacy primarily depends on the spatial structure of the forcings and the way they project onto the various different feedback mechanisms. Therefore, different patterns of RF and any non-linearities in the forcing response relationship affects the efficacy. Nearly all studies that examine it find that high-latitude forcings have higher efficacies than tropical forcings. Efficacy has also been shown to vary with the vertical distribution of an applied forcing.

Forcings that predominately affect the upper troposphere are often found to have smaller efficacies compared to those that affect the surface. However, this is not ubiquitous as climate feedbacks will depend on the static stability of the troposphere and hence the sign of the temperature change in the upper troposphere.

Long-Lived Greenhouse Gases

The few models that have examined efficacy for combined LLGHG changes generally find efficacies slightly higher than 1.0. Further, the most recent result from the NCAR Community Climate Model GCM indicates an efficacy of over 1.2 with no clear reason of why this changed from earlier versions of the same model. Individual LLGHG efficacies have only been analysed in two or three models. Two GCMs suggest higher efficacies from individual components. In contrast another GCM gives efficacies for CFCs and CH4 that are slightly less than one. Overall there is medium confidence that the observed changes in the combined LLGHG changes have an efficacy close to 1.0 but there are not enough studies to constrain the efficacies for individual species.

Solar

Solar changes, compared to CO2, have less high-latitude RF and more of the RF realised at the surface. Established but incomplete knowledge suggests that there is partial compensation between these effects, at least in some models, which leads to solar efficacies close to 1.0. All models with a positive solar RF find efficacies of 1.0 or smaller. One study finds a smaller

Effects of Global Warming on Ecosystems

efficacy than other models. However, their unique methodology for calculating climate sensitivity has large uncertainties.

These studies have only examined solar RF from total solar irradiance change; any indirect solar effects are not included in this efficacy estimate. Overall, there is medium confidence that the direct solar efficacy is within the 0.7 to 1.0 range.

Ozone

Stratospheric ozone efficacies have normally been calculated from idealised ozone increases. Experiments with three models found higher efficacies for such changes; these were due to larger than otherwise tropical tropopause temperature changes which led to a positive stratospheric water vapour feedback. However, this mechanism may not operate in the two versions of the GISS model, which found smaller efficacies. Only one study has used realistic stratospheric ozone changes; thus, knowledge is still incomplete. Conclusions are only drawn from the idealised studies where there is medium confidence that the efficacy is within a 0.5 to 2.0 range and established but incomplete physical understanding of how and why the efficacy could be larger than 1.0. There is medium confidence that for realistic tropospheric ozone perturbations the efficacy is within the 0.6 to 1.1 range.

Scattering Aerosol

For idealised global perturbations, the efficacy for the direct effect of scattering aerosol is very similar to that for changes in the solar constant. As for ozone, realistic perturbations of scattering aerosol exhibit larger changes at higher latitudes and thus have a higher efficacy than solar changes. Although the number of modelling results is limited, it is expected that efficacies would be similar to other solar effects; thus there is medium confidence that efficacies for scattering aerosol would be in the 0.7 to 1.1 range. Efficacies are likely to be similar for scattering aerosol in the troposphere and stratosphere. The efficacy of the cloud albedo RF accounts for cloud lifetime effects. Only two studies contained enough information to calculate efficacy in this way and both found efficacies higher than 1.0. However, the uncertainties in quantifying the cloud lifetime effect make this efficacy very uncertain. If cloud lifetime effects were excluded from the efficacy term, the cloud albedo efficacy would very likely be similar to that of the direct effect.

Absorbing Aerosol

For absorbing aerosols, the simple ideas of a linear forcingresponse relationship and efficacy can break down. Aerosols within a particular range of single scattering albedos have negative RFs but induce a global mean warming, that is, the efficacy can be negative. The surface albedo and height of the aerosol layer relative to the cloud also affects this relationship.

Studies that increase BC in the planetary boundary layer find efficacies much larger than 1.0. These studies also find that efficacies are considerably smaller than 1.0 when BC aerosol is changed the boundary layer. These changes in efficacy are at least partly attributable to a semi-direct effect whereby absorbing aerosol modifies the background temperature profile and tropospheric cloud. Another possible feedback mechanism is the modification of snow albedo by BC aerosol; however, this report does not classify this as part of the response, but rather as a separate RF.

Most GCMs likely have some representation of the semi-direct effect but its magnitude is very uncertain and dependent on aspects of cloud parametrizations within GCMs. Two studies using realistic vertical and horizontal distributions of BC find that overall the efficacy is around 0.7. However, Hansen acknowledge that they may have underestimated BC within the boundary layer and another study with realistic vertical distribution of BC changes finds an efficacy of 1.3. Further, Penner also modelled BC changes and found efficacies very much larger and very much smaller than 1.0 for biomass and fossil fuel carbon, respectively found similar efficacies for biomass and fossil fuel carbon). In summary there is no consensus as to BC efficacy and this may represent problems with the stratospherically adjusted definition of RF.

Other Forcing Agents

Efficacies for some other effects have been evaluated by one or two modelling groups. Hansen found that land use albedo RF had an efficacy of roughly 1.0, while the BCsnow albedo RF had an efficacy of 1.7. Ponater found an efficacy of 0.6 for contrail RF and this agrees with a suggestion from Hansen that high-cloud changes should have smaller efficacies. The results of Hansen and Forster and Shine suggest that stratospheric water vapour efficacies are roughly one.

EFFICACY AND THE FORCING-RESPONSE RELATIONSHIP

Efficacy is a new concept introduced since the TAR and its physical understanding is becoming established. When employing the stratospherically adjusted RF, there is medium confidence that efficacies are within the 0.75 to 1.25 range for most realistic RF mechanisms aside from aerosol and stratospheric ozone changes. There is medium confidence that realistic aerosol and ozone changes have efficacies within the 0.5 to 2.0 range. Further, zero-surface-temperature-change RFs are very likely to have efficacies significantly closer to 1.0 for all mechanisms.

BASIC COMPONENTS OF THE ECOSYSTEM

It is helpful to visualize ecosystems as consisting of four basic components: abiotic substances, producer organisms, consumer organisms and decomposer organisms.

Abiotic Components

Abiotic substances are the inorganic and organic substances not momentarily present in living organisms. These include water, carbon dioxide, oxygen, nitrogen, minerals, salts, acids, bases and the entire range of elements and compounds outside living organisms at any given point in time. Many elements may be tightly bound in inorganic compounds, such as silicon in sandstone or aluminum in feldspar, and are unavailable to living organisms. Elements which are readily available to living organisms such as free O_2 or CO_2, or they may be in an inaccessible form such as silicon dioxide (SiO_2) in quartz, a major component of granite. Similarly, potassium may be readily available to plants in the form of KCL in soil, but relatively unavailable in the form of $KAlSi_3O_8$ in orthoclase or monoclinic feldspar, one of the commonest of all minerals.

An important property of an ecosystem which determines its productivity is the form and composition in which bioactive elements and compounds occur. For example, an ecosystem may have a substantial abundance of vital nutrients, such as nitrates and phosphates, but if they are present in relatively insoluble particulate form as they would be if linked to ferric ions, they would not be so readily available to plants as if they were in the soluble form of potassium or calcium nitrate and phosphate. One of the most important qualities of an ecosystem is the rate of release of nutrients from solids, for this regulates the rate of function of the entire system.

Producers

Producer organisms are bacteria and plants which synthesize organic compounds. They are said to be autotrophic or self-productive, in that they take inorganic compounds and manufacture organic materials and living protoplasm from them. All green plants, including microscopic algae, are producer organisms since they exhibit photosynthesis, and some bacteria are producers since they may exhibit chemosynthesis or photosynthesis. Obviously, all life depends upon the basic productive capacity of green plants and bacteria.

Consumers

Consumer organisms are animals which utilize the organic materials directly or indirectly manufactured by plants. Consumers are unable to produce their own organic compounds for basic nutritive purposes. They are said to be heterotrophic, which means different or varied in nutritional source. Primary consumers or herbivores directly consume the organic compounds of plants. Secondary consumers may be omnivores or carnivores which depend partially or entirely on other animals for food. Tertiary and quartenary consumers may be the second or third-stage predator, for example, a hawk feeding on a weasel which in turn consumed a mouse.

Decomposers

Decomposer organisms are bacteria and fungi which degrade organic compounds. Their nutrition is said to be saprophytic, that is, associated with rotten an ecosystem—they reduce the complex organic molecules of dead plants and animals to simpler organic compounds which can be absorbed by green plants as vital nutrients. They provide the final essential link in the cycle of life. They are necessary for the renewal of life, for if decomposers were not active, organic compounds would become locked into complex insoluble molecules which could not be utilized as nutrients by plants.

Ecosystems involve, of course, a wide variety of life forms not specifically mentioned in the preceding paragraphs, but virtually all components of an ecosystem can be classified into producers, consumers, or decomposers. For example, parasites are merely specialized consumers. Plant parasites feed directly on plants and are thus herbivores; animal parasites derive their nutrition from other animals, and are thus carnivores differing from predators only in the fact that they normally do not kill the host. Scavengers such as vultures are also carnivores, differing from predators by the fact that they feed on an animal after it has died from some other cause.

INCOMPLETE ECOSYSTEMS

Almost all ecosystems have all four basic components discussed above, though in some cases it is possible for incomplete ecosystems to exist. These are ecosystems lacking one or more basic components.

An example of an incomplete ecosystem lacking producers is the abyssal depths of the sea where only consumers and decomposers exist. In the realm of complete darkness green plants cannot survive. Scavengers and decomposers live on the fall-out of animals, plants and organ matter from the upper layers of the ocean. Predators might also be present to feed upon the scavengers. Hence, the ecosystem depends on extrinsic production, namely, the fall-out from upper levels. It might be possible, of course, for a few chemosynthetic bacteria to be present, but they would not produce a significant volume of organic material.

The same situation exists in caves where complete darkness prevents the growth of green plants. Again, a few chemosynthetic bacteria might be present, but they would not produce a significant amount of organic material. Practically all cave-dwelling-animals must depart from the cave, as do bats, or depend on extrinsically produced nutrients which enter the cave by flowing water or seepage.

The central core of the city might also be considered an incomplete ecosystem without producers, at least from the human standpoint. Some green plants obviously exist, but they would not supply meaning ful production for humans, sparrows, dogs, cats, etc. For all of these, the inner city requires extrinsic production and imported food. The only other alternative is for the inhabitants to leave the inner city and feed in peripheral areas. This undoubtedly occurs with starlings and pigeons, creating an ecological situation analogous to bats in a cave. The lack of production in an inner city is not due to a lack of light, but to a lack of soil and suitable substrate.

In other w ays, cities may be considered incomplete ecosystems, ecologically parasitic upon the surrounding landscape. Not only do they import food, but they must also import fresh air and water. At the same time, they must export waste products—sewage, solid waste, carbon dioxide, sulphar dioxide, etc. If cities were encapsulated from their surrounding environments they would soon perish from thirst, starvation, asphyxiation, or the accumulation of waste products. In exchange for this life support, cities, of course, provide a great many economic and cultural benefits—jobs, housing, transportation, manufacturing, education, etc. So the relationship between city and landscape is vital in both directions, but it is particularly important to remember, as cities expand, that they cannot sustain themselves.

Incomplete ecosystems also exist in specialized cases where producers and decomposers are present without consumers. A theoretical example would be a massive bloom of some toxic algae in an aquatic ecosystem, where the algae would create toxic conditions for zooplankton and fish and all other possible consumers. Then a process of excessive production and massive decomposition would go hand in hand. This would be a highly unstable and undesirable circumstance, but it has been known to occur, as, for example, in the red tides of Florida. A third type of incomplete ecosystem might even be called an abiotic ecosystem, that is, one without living organisms, a self-contradiction in terms. They should more properly be called abiotic environments. Apollo space flights have shown so far that the moon is abiotic. Local areas on earth may be abiotic, for example, the high altitude ice plateau of Antarctica is probably devoid of living organisms over rather extensive areas. Closer home for most of us, the Copperhill basin of Tennessee is an area devoid of life, where fumes from copper smelters are toxic to all organisms within a certain downwind area. Possibly a few bacteria exist in very limited places, but for all practical purposes no plants or animals can survive.

GLOBAL WARMING STRESS TO CORAL REEFS

The World Wildlife Fund (WWF) reports that coral reefs around the world have been severely damaged by unusually warm ocean waters. They predict that less than 5 per cent of Australia's Great Barrier Reef will remain by 2050 if the world fails to reduce CO_2 emissions. They project that "if the present rate of destruction continues, a good proportion of the world's coral reefs could be killed within our lifetime." WWF has identified both the Seychelles Islands and American Samoa as locations under high stress for coral bleaching. The United Nations Educational, Scientific and Cultural Organization (UNESCO) recognizes the Seychelles Islands as a natural World Heritage Site.

They have a high diversity of coral and support rare land species, such as the giant tortoise. In addition to increasing ocean temperatures, these areas are also threatened by global warming because of more frequent tropical storms (which could break up the coral) and more frequent rains, flooding, and river run-off (which deposits sediments in the ocean). The National Aeronautics and Space Administration (NASA), reef habitats are so complex and warrant so much exploration, "that for marine biologists, the destruction of the reefs has proven to be as frustrating as it is heart breaking. At the rate the reefs are disappearing, they may be beyond repair by the time a comprehensive plan to save reefs can be put into place." Scientists at NASA's Goddard Space Flight Centre and at several universities around the world have at least a partial solution. They have been interpreting satellite imagery of the world's oceans obtained from Landsat 7 and other high-resolution sensors to map the shallow waters around the ocean's margins, creating maps of the reef environments.

With funding, they expect to soon have a comprehensive map of the world's reef systems that they can use to identify large-scale threats to the reefs. In the meantime, Abdul Azeez Abdul Hakeem, a Maldivian scientist, is dedicated to helping corals on Maldives—a nation of 1,200 islands in the Indian Ocean—survive global warming. Azeez is the director of conservation for an eco-friendly resort called Banyan Tree Maldives. Besides being a premier vacation resort, it also hosts a worldclass marine laboratory managed by Azeez. The staff spends most of their time studying the coral and maintaining the ecosystem—the reefs provide homes for many exotic fish and attract travellers from all over the world. Azeez became involved with coral in 1998 when a strong El Niño warmed the ocean and put the coral at risk. "It rose to about 91°F (33°C), so 33 was boiling hell, and about 80 to 90 per cent of the corals in the Maldives died. I never believed that an entire region could be wiped out. No one believed that this could happen until it hit us. Then only I also realised yes we are in danger because of global warming and this can happen again and again." When that happened, Azeez began looking for ways to protect the coral.

He also knew that corals on an artificial reef nearby had survived the 1998 El Niño. That reef was an experimental design that used electricity. At that time, no one knew how it had protected the reef from the extreme heat. Azeez took the design idea and built an electric reef from steel bars and wired it to a power source on the beach at the Banyan Tree Resort. The small electric current caused the minerals from ocean water to build up on the steel, forming a thick limestone crust that is perfect for coral. The electric reef was situated on the far side of the island, submerged in 15 feet (5 m) of water.

The reef is maintained as a type of underwater topiary. An immense variety of corals grow on it like an underwater garden; all thriving in their sheltered habitat. "Finger corals, hard corals, massive corals. We have tried to plant as many species as we can," Azeez explains. "The reef also attracts both predators and prey, just as a healthy reef ecosystem does." Azeez considers his electric reef as a type of "greenhouse for corals." He believes it will keep a critical mass alive when the next El Niño strikes. Azeez claims, "You can take pieces from the corals on this structure to that one and make your own garden again." This transplant process seems to be working for Azeez in the Maldives. "Our enemy, the real threat," says Azeez, "is global warming."

In a report in National Geographic News, in 1995, it was stated that roughly 10 per cent of the coral colonies in Belize had died. "This coral bleaching is pretty solidly tied to rising ocean temperatures," said Melanie McField, a Belize-based reef scientist with the WWF. "It's a fact that global temperatures have risen. There's lots of data and little argument that increased ocean temperatures are the primary agent of bleaching.

As for tying overall temperature increases to overall global warming, there is still some debate, but less every year. I think the majority of scientists agree that global warming is happening and that it's the root cause of these coral bleaching events. Rather than throwing up our hands and saying 'we can't control that,' we've got to be even more diligent and try even harder to control local impacts such as pollution and overfishing," Mc Field said. Even in view of the unstoppable damage to reefs in the future, conservationists say that given the complex factors affecting coral health, there is still a lot that can be done to help reefs recover if action is taken now.

Protection and Conservation

As the world becomes more aware of the perilous condition of the fragile marine habitat, concern is growing over the future condition and existence of these beautiful areas. As more is learned about what is happening to oceanic ecosystems, more attention is being given to the creation of protected marine areas in order to conserve, manage, and protect ocean resources so that they will not become extinct. The location of conservation areas needs to be carefully chosen to be able to protect and provide for the greatest diversity of species. On June 16, 2006, former president George W. Bush designated the Northwest Hawaiian Islands a national monument.

This created the largest marine protected area in the world, providing important habitat for thousands of marine species that rely on it for survival. Steve McCormick, past president of the Nature Conservancy, "We commend the administration for its foresight and leadership in protecting this incredible area. Designating the northwest Hawaiian Islands as a national monument will ensure that this national treasure will remain healthy and intact for generations to come." This preservation area includes 5,019 square miles (13,000 km2) of coral reefs, comprising 70 per cent of all coral reefs in the United States. The National Geographic News, a consortium of Latin American nations, conservation groups, and United Nations agencies are creating one of the world's largest marine protected areas to be known as the Eastern Tropical Pacific Seascape. The new reserve will cover 521 million acres (211 million ha) of ocean from Costa Rica's Cocos Island to Ecuador's Galápagos Islands. The purpose of the preserve is to protect a wide range of ocean species, expand existing marine reserves, and consolidate current and planned conservation efforts. It will be the largest marine conservation area in the Western Hemisphere. The region already has these currently listed World Heritage Sites: Galápagos Islands and Marine Reserve (in Ecuador), Cocos Island (Costa Rica), and Coiba National Park (Panama). There are several conservation organizations in existence today that are geared towards helping promote healthy ocean ecosystems. Fortunately for the environment, as well as future generations, tens of thousands of people choose to get involved and make a difference.

THE EFFECTS OF GLOBAL WARMING ON OPEN OCEANS

Because the oceans are so vast, much of the predictions made about the Earth's open oceans in light of global warming have been generated by

Effects of Global Warming on Ecosystems

computer models. Because the two most important functions that govern the behaviour of the ocean are temperature and circulation, these calculations are fairly straightforward in computer models and readily calculable. Another thing that makes it possible to model the oceans using computers is that humans have not had as large an impact on the oceans as they have on land.

The Thermohaline Circulation, the Great Conveyor Belt

Because of the interactions of temperature and salinity in the world's oceans, movement is generated that links them all. This linking of the oceans allows heat energy to be transported globally. Cold water is much denser than warm water, so it sinks to the bottom of the ocean while the warmer water rides above it. Water that is high in salt content is also denser, which forces it below water that is lower in salinity. These differences in density are what generate the movement of the oceans. The most important movement is the thermohaline (thermo for "temperature" and haline for "salt content") system, commonly known as the great ocean conveyor belt. This system moves cold dense seawater from the North Atlantic Ocean surface to deep water and then continues through the Indian and Pacific Oceans and back to the Atlantic.

This process takes centuries for an entire cycle to be completed, and it is one of the most important currents in the ocean for regulating heat on the Earth's continents. This is the current that picks up heat in the equatorial region, rises to the ocean's surface, and travels northward via the Gulf Stream/North Atlantic Current past the west coast of Europe, moderating the climate there, making it much milder than it otherwise would be at its northern latitude. As the current reaches its northern limit in the Arctic, it cools and sinks and takes with it the nutrients as well as oxygen and carbon dioxide (CO_2) that it absorbed while it was at the surface of the ocean, creating a sink for CO_2. Deep-sea organisms are able to use the nutrients and oxygen, but the CO_2 does not get used by plants in photosynthesis because the ocean depths are too dark. If global warming intensifies and freshwater from melting glaciers and ice caps is added to the Arctic waters, it will dilute the salinity and slow or stop the vertical mixing between the ocean surface and the deep sea. The result of this would be to slow, or even halt, the conveyor belt.

If this were to happen, warmth would no longer be delivered to Europe via the Gulf Stream, which could put Europe into an ice age. There would be many other effects as well. Diminished vertical mixing between the ocean's surface and depths would reduce the upwelling in temperate and subtropical latitudes. Oxygen would not be effectively transported from the surface to the ocean depths. Over time—centuries— the deep ocean waters would become devoid of oxygen, and they would either become hypoxic (low in oxygen concentrations) or anoxic (completely lacking in oxygen). The dispersal of nutrients from the surface to the depths would also be impaired, leaving deep-sea animals, including commercial fish and squid, without nutrition. This could destroy the fishing industry. As far as the present rate of CO_2 uptake by the oceans, the Pew Climate Centre believes there is disagreement as to whether CO_2 would be less likely to be absorbed if the ocean depths were not mixing effectively, removing the ocean as a CO_2 sink. This would increase CO_2 concentrations in the atmosphere or it would cause an increase in biological productivity in the upper ocean layers and increase the CO_2 uptake.

The IPCC in their third report, general circulation models (GCMs) predict a wide range of global ocean circulation responses to global warming, ranging from "no response" to a "40 per cent decline in circulation before 2100." Dr. Thomas F. Stocker of the University of Bern's Physics Institute, "changes in the thermohaline circulation are likely with global warming, but the extent of the changes are uncertain at this point." Dr. Peter Robert Gent, chairman of the science steering committee for the Community Climate System Model, working with the National Science Foundation and the U.S. Department of Energy, says he "does not expect a collapse of the thermohaline circulation during the 21st century because of the feedbacks that help stabilize circulation in response to warming."

He warns, however, that "presently, little is known about the feedback mechanisms that affect the thermohaline circulation pattern." Another important consideration, he says, is "that these projections are made under the assumption that atmospheric CO_2 levels do not exceed a doubling of preindustrial levels before 2100. Models that were run basing a CO_2 level above double the preindustrial levels before 2100 did simulate the collapse of the thermohaline circulation by 2100—which could put Europe in an ice age." These are still major unknowns concerning global warming and much more research needs to be done.

Ocean Acidification

Ocean acidification is the ongoing decrease in the pH value of the Earth's oceans caused by their steady uptake of anthropogenic (human-caused) CO_2 from the atmosphere. The German Advisory Council on Global Change, the oceans presently store about 50 times more CO_2 than the atmosphere and 20 times more than the terrestrial biosphere and soils.

The ocean is not only an important CO_2 reservoir, it is also the most important long-term CO_2 sink. Due to pressure differences between the atmosphere and seawater, some of the anthropogenic CO2 dissolves in the surface layer of the ocean. Over periods of time—decades to centuries to millennia—the CO2 is carried into the deep sea by ocean currents. Christopher L. Sabine of NOAA's Pacific Marine Environmental Laboratory, the ocean is presently taking up 2 Gt (gigatons) of carbon annually, which just as to the IPCC is the equivalent of roughly 30 per cent of the anthropogenic CO2 emissions. (A gigaton is one billion tons [900,000,000 metric tons].) The IPCC calculates that between 1800 and 1995, the oceans absorbed about 118 Gt of carbon.

This corresponds to about 48 per cent of the cumulative CO2 emissions from fossil fuels, or 27–34 per cent of the total anthropogenic CO2 emissions, including those from land-use changes such as deforestation. Currently, anthropogenic CO2 can be tracked to a depth of 3,281 feet (1,000 m) in the ocean.

It has not sunk to the bottom of the ocean yet because the vertical mixing process is so slow that it takes a long time. The North Atlantic region is different, however. Because the vertical mixing happens much more readily with the currents there, anthropogenic CO2 has been traced to a depth of 9,843 feet (3,000 m). In the ocean, CO2 behaves differently than it does in the atmosphere. In the atmosphere, it is chemically neutral; in the ocean it is chemically active. Dissolved CO2 contributes to the reduction of the pH value and causes an acidification of seawater, which is a property that can be easily measured. Since the onset of the Industrial Revolution in the 1700s, the pH value of the oceans has dropped (become more acidic) by about 0.11 units.

Starting from a slightly alkaline, preindustrial pH value of 8.18, the acidity of the ocean has increased at the surface. Based on modeling done by the IPCC, if atmospheric CO2 concentrations reach 650 parts per million (ppm) by the year 2100, a decrease in the average pH value by 0.30 units can be expected compared to preindustrial levels. If concentrations increase to 970 ppm, the pH level would drop by 0.46 units. The scale of changes will vary regionally, which will affect the magnitude of any biological effects. Based on the following data from the Royal Society in London, they believe that ocean acidification ... is essentially irreversible during our lifetimes. It will take tens of thousands of years for ocean chemistry to return to a condition similar to that occurring at preindustrial times (about 200 years ago). Our ability to reduce ocean acidification through artificial methods such as through the addition of chemicals is unproven. These techniques will at best be effective only at a very local scale, and could also cause damage to the marine environment. Reducing CO2 emissions to the atmosphere appears to be the only practical way to minimize the risk of large-scale and long-term changes to the oceans.

All the evidence collected to date points to ocean acidification being caused by human activity—burning fossil fuels, deforestation, and land-use change.

The magnitude of acidification can be predicted accurately. What cannot be predicted yet is the physiological effects on various organisms. Headway has been made, however, on discovering the ramifications of acidification on the process of calcification—the process by which organisms such as corals and mollusks make shells and plates from calcium carbonate. The tropical and subtropical corals are expected to be the worst affected. This could destroy coral reef habitats. Phytoplankton and zooplankton, which are major food sources for fish and other animals, will also be negatively affected. This is one area of global warming where research is still in its infancy and needs much more to be done in order to understand both the short- and long-term effects.

SEAWEED—A CARBON SINK

Recent studies indicate that seaweed may be capable of sucking CO_2 out of the atmosphere at rates comparable to the world's rain forests. Chung Ik-Kyo, a South Korean environmental scientist, "The ocean's role is neglected because we can't see the vegetation. But under the sea, there is a lot of seaweed and sea grass that can take up carbon dioxide." This concept has such potential, in fact, that 12 Asian-Pacific countries are currently working together to calculate how much CO_2 is being absorbed from the atmosphere by plants and to increase carbon storage through carbon sinks.

In 2007, the United Nations Climate Change Conference in Bali brought together 10,000 participants from 180 countries and adopted the Bali Road Map, various decisions towards reaching a secure climate future. A major issue addressed was identifying those things that can be potentially used as carbon sinks to remove carbon from the air. One possible sink, just as to Chung Ik-Kyo, is seaweed.

There is a tremendous amount of seaweed produced and available each year—8 million tons (7.3 million metric tons)—harvested from wild or cultivated sources.

This theory has met with criticism, however. Some believe it will be a challenge to keep the carbon, once it is absorbed in the seaweed, from reentering the atmosphere at some point. Another negative side effect is the uncertainty about exactly how an increase in seaweed production would affect ocean navigation or fisheries.

The world's current biggest producers of seaweed are China, South Korea, and Japan. Those that favour the plan of using seaweed say that seaweed and algae's rapid rate of photosynthesis (the process of turning CO2 and sunlight into energy and oxygen) make them the perfect candidates for absorbing carbon.

Lee Jae-young of South Korea's fisheries ministry, "Seaweeds can absorb five times more carbon dioxide than plants on land." John Beardall with Australia's Monash University says, "The oceans account for 50 per cent of all the photosynthesis on the Earth.

These are very productive ecosystems." In addition to using seaweed as a CO2 storage, these same researchers have also suggested that it be used to produce clean-burning biofuels. Skeptics of this concept say that the reason trees are so effective at carbon storage is because they live for many years. Seaweed is grown and harvested in terms of a few months.

They feel that this would make this type of carbon storage difficult to measure and control. Other critics wonder if removing water from the seaweed as it is converted to fuel would require a larger input of energy than that which would be received, making it environmentally unwise. Chung responds by saying, "The idea is still in its infancy. In terms of ball games, we are just in the bullpen, not the main game yet." Currently, South Korea and Japan are the leaders in this type of research.

THE EFFECTS OF GLOBAL WARMING ON COASTAL LOCATIONS

The Pew Centre on Global Climate Change is a non-profit organization that brings together business leaders, policy makers, scientists, and other experts worldwide to create a new approach to managing the problem of global warming-what they refer to as an extremely"controversial issue." Not slanted in any particular way politically or economically, they"approach the issue objectively and base their research and conclusions on sound science, straight talk, and a belief that experts worldwide can work together to protect the climate while sustaining economic growth." The Pew Centre was established in 1998 and has"issued more than 100 reports from top-tier researchers on key climate topics such as economic and environmental impacts and practical domestic and international policy solutions."

The centre's leaders and staff hold briefings with members of Congress and administration officials in the U.S. government, as well as with leaders of international governments. They also work with 43 major corporations in business roles to encourage them to help promote practical solutions to solving the global warming crisis. In January 2007, the Centre was one of the inaugural members of the U.S. Climate Action Partnership-an alliance of major businesses and environmental groups that calls on the federal government to enact legislation requiring significant reductions of greenhouse gas emissions. The Pew Centre on Global Climate Change, in terms of global warming, the biggest impact on estuarine and marine systems will be temperature change, sea-level rise, the availability of water from precipitation and run-off, wind patterns, and storminess. In these oftenfragile systems, temperature has a direct and serious effect. For the sea life living within the ocean, temperature directly affects an organism's biology, such as birth, reproduction, growth, behaviour, and death. The remainder of this part will further explore the specific effects global warming will have on temperature, sea-level rise, wind circulation, aquaculture, algae blooms and disease, and the long-term effects of coastal building.

TEMPERATURE

Temperature extremes (both high and low) can be deadly to living organisms. Many species are vulnerable to temperatures just a few degrees higher than what they are accustomed to. Even increases in temperature as little as 1.7°F (1°C) can seriously harm certain species. For example, from 1976- 77, the species of reef fish off Los Angeles diminished by 15-25 per cent when temperatures abruptly warmed 1.7°F (1°C). Even if the temperature is not high enough to kill, it can be high enough to negatively influence the organism's life functions such as metabolism, growth, behaviour, and physiological factors such as the timing of reproduction rates of egg and larval development.

Temperature also influences where organisms can survive and controls how large a given population can become.

One area where this has long been documented is along the West Coast of North America. The nearshore sea surface is predominantly warm and the offshore area is predominantly cool, creating a region that has long been famous as a rich, diverse fishery. Temperature differences can also influence interaction between species, such as predator-prey, parasite-host, and other relationships that may develop over the struggle for limited resources. If temperatures change the distribution patterns of organisms, it could also change the balance of predators, prey, parasites, and competitors in an ecosystem, completely readjusting balances, food chains, behaviours, and the equilibrium of the ecosystem. Global warming can also change the way that species interact by changing the timing of physiological events. One of the key changes that it could alter is the timing of reproduction for many species. Rising temperature could interfere with the timing of birth being correlated with food availability of that species.

Effects of Global Warming on Ecosystems

This can be a problem, for example, for bird species that migrate and depend on a specific food source to be available when they reach their breeding grounds. If warmer temperatures have changed the timing and it is a few weeks off of when the food will be available and no longer synchronized with migrating birds, it could leave the birds without available food, threatening their survival. Temperature also plays an important role with oxygen because it directly influences the amount that water can hold. The warmer the water, the less oxygen it holds. As global warming continues to force temperatures to rise, less oxygen will be available for marine species, which could then threaten their survival. A significant amount of data has already been collected on these variables. The Pew Centre, areas that still need much more data collected and research applied concern the ocean temperatures influence on the interactions among organisms, such as predator-prey relationships, parasite-host relationships, and the interplay of species concerning competition for resources.

It is also possible to model the effects of sea-level rise in shallow continental margins, such as flooding wetlands and shoreline erosion. What is more difficult to predict and needs more research are the global warming effects on precipitation amounts, wind patterns, and intense weather. Precipitation directly affects estuaries because it affects the run-off into estuaries, which influences estuarine circulation.

Sea-Level Rise

Sea-level rise will mount from the melting of glacial and polar land ice. The Pew Centre, the effects of sea-level rise will vary by location, how fast the sea level rises, and the biogeochemical responses of the individual ecosystems involved. One of the areas identified as being the most susceptible to damage from sea-level rise is the low-lying, flat wetlands located in the middle and south Atlantic and the Gulf of Mexico (involving states such as Alabama, Florida, Mississippi, Georgia, and Texas). A wetland is an area on shore that has a wet, spongy soil. These areas are also referred to as swamps, marshes, and bogs. As sea levels rise, ocean water will submerge and erode the shorelines. In the natural areas covered with marshes and mangroves (common to Florida and the Gulf states), sea levels will flood the wetlands and waterlog the soils.

Because the plants that live in the wetlands, which do not contain salty water, are not accustomed to salt, the salty ocean water will kill them. Because wetlands provide habitat for wildlife, including several migrating birds, this would also destroy their habitat. Based on research by the Pew Centre, if the wetlands located along the Gulf of Mexico are not bordered by human development (such as homes) and the inland slope is relatively flat, and the sea-level rise is gradual, they would have a chance of surviving by migrating further inland to escape the inundation of salt water. The Pew Centre states that "these areas would be safe at least at the rates of rise currently projected for the next 50 to 100 years." The areas that would be hard hit are those that cannot migrate inland because urbanization has grown right up to their shoreline, effectively removing any potential wetland habitat. This is very detrimental to the environment because wetlands are an important part of the biological productivity of coastal systems. Marshes provide many critical services; they function not only as habitats for wildlife, but as nurseries for breeding and raising young and as refuges from predators. Wetlands function as part of an integrated system.

If they are jeopardized, their loss will affect the availability and transfer of nutrients, the flow of energy, and the availability of natural habitat needed by multitudes of organisms already living there. One of the most unfortunate losses will be those areas where rare, threatened, or endangered plant and animal species live, such as the American alligator, Florida black bear, West Indian manatee, Florida panther, southern bald eagle, snowy egret, and roseate spoonbill. If invading salt water destroys these habitats, all these species could become extinct.

Wind Circulation

Another consideration in the marine environment is wind circulation. Winds are created by the uneven heating at the Earth's surface, specifically the high thermal gradients between the equator and the poles. Under global warming conditions, however, the polar regions will be subjected to higher temperatures, thereby reducing the temperature gradient between the equator and the poles. This could cause a weakening in the overall wind circulation around the Earth. The winds drive the surface currents of the Earth's oceans.

A slowdown of the ocean surface circulation could negatively affect the structure and function of both the open ocean and the near shore ecosystems. Weak currents would make it difficult for any currents to transport nutrients where they need to be deposited. If wind speed and direction are altered, which is what experts at the Pew Centre have predicted will occur with global warming, the productivity of estuarine and marine systems will also be affected.

One of the reasons why some coastal areas are so rich and productive in fishery resources, such as the West Coast of the United States, is because of the coldwater upwelling that occurs along the coastal margins. The thermal properties and circulation patterns function to bring the cold water from the deep ocean waters to the surface along the coastlines, supplying the coastal fisheries with an abundance of fish. Nutrients that are delivered from these deeper waters to the surface help provide for the phytoplankton along the coasts. The upwelling process is controlled by the winds that blow along shore. Upwelling can be minimized, however, if the water column is stratified by differences in temperature or salinity, which act as a barrier to upward movement through the water column. The effects of this type of scenario were played out off the coast of California from 1951 to 1993. The sea surface warmed during this period 2.5°F (1.5°C), resulting in a decreased upwelling of nutrientrich cold waters. This caused a 70 per cent decline in the abundance of zooplankton, which in turn harmed the coastal food webs, negatively affecting fish and birds. There are different opinions on this matter, however. Andrew Bakum of the National Marine Fisheries Service of the National Oceanic and Atmospheric Administration (NOAA), his research leads him to believe that global warming may actually increase coastal upwelling in some areas.

He believes this is possible because most greenhouse gas models project more warming over land than over open oceans, which would increase the temperature contrast between the land and the ocean. This resultant contrast would strengthen the low-pressure cells that usually form over land that are adjacent to offshore high-pressure cells. The relationship of forces between these two pressure cells would create alongshore winds that would create a strong upwelling. On the East Coast of the United States, circulation patterns could slow down transport of nutrients and fish species, such as blue crabs and bluefish, which would lower the abundances of these particular species along the coasts and within estuaries. When their populations are decreased, the coastal ecosystem becomes threatened.

Aquaculture

Coastal aquaculture, the farming of fish from the ocean in special facilities, has become a key food source in recent years. The Food and Agriculture Organization (FAO), global aquaculture production has been increasing since the 1950s at a growth rate of about 10 per cent per year since 1990. By 2030, the FAO projects that aquaculture harvests will be greater than capture harvests. Marine aquaculture is a steadily growing industry. The effect that global warming will have on aquaculture is mixed. On one hand, the higher temperatures could enhance growth rates of aquaculture species and could make it possible to run aquaculture operations in locations that are currently too cold. On the other hand, areas that are suitable now for aquaculture may become too warm for existing species to survive.

The Pew Centre's research suggests that as conditions change, if the temperatures rise slowly enough, aquaculture operations should be able to keep up with the changes and keep negative effects to a minimum.

Algae Blooms and Disease

One major drawback of aquaculture facilities is that they affect local ecosystems when their concentrated wastes pollute the surrounding environment, encouraging algae blooms. If global warming becomes an issue, algae blooms will become more common. Warm water holds less oxygen and accelerates the microbial decomposition of the aquaculture wastes. Oxygen concentrations are further lowered, exacerbating the problem. As global warming continues and temperatures climb upward, disease will become more prevalent, distributing the pathogens to more areas. Not only does this affect ocean life, it also affects humans. An example of this has been showed with oysters.

The protozoan Perkinsus marinus (Dermo) is a pathogen that threatens the health of oysters. Cool ocean temperatures keep infection by this pathogen to a minimum. As water temperatures have risen in recent years, the spread of this pathogen has been documented spreading to the northeastern states of the United States. Based on this evidence, Dr. T. Cook of Rutgers University's Institute of Marine and Coastal Sciences says longterm climate changes may produce shifts in salinity and temperature that enable pathogens to spread. Climate change could also affect the distribution of other aquatic diseases in a similar manner. Warmer coastal waters along with eutrophication (an increase in chemical nutrients—usually compounds containing nitrogen or phosphorus—in an ecosystem) can also increase the intensity of harmful algal blooms, which can destroy habitat and shellfish nurseries and are also toxic to both marine species and humans. Unfortunately, these blooms have been recently increasing worldwide, possibly because they are correlated to global warming.

One of the concerns is that consumption of shellfish that have ingested harmful algae can cause neurotoxic poisoning in humans. The Pew Centre, there are two schools of thought concerning species adaptation and survival in marine ecosystems. One group thinks that marine ecosystems will have far fewer extinctions than terrestrial ecosystems because marine species have wider geographic temperature ranges and the ability and opportunity to migrate to new habitats. The other group thinks that the lack of evidence of recent marine extinctions is simply because the ocean systems are so vast that scientists are not aware of them, and there is not enough data collected about the oceans.

Long-term Effects of Coastal Building

The New York Times on July 25, 2006, entitled "Climate Experts Warn of More Coastal Building," 10 climate experts from around the United States say that the unchecked coastal development in geographic locations that are vulnerable to hurricanes and other forms of severe weather, along with the lack of government regulation on development, is the biggest reason why extreme, violent weather is currently causing so much human suffering, loss of life, and property damage. The substance states that the opinion of the experts is that: "Whatever the relationship between hurricanes and climate, hurricanes are hitting the coasts and houses should not be built in their path." One of the main problems is social/political in nature.

Because coastal areas are popular places to live, pressures are put on states and Congress to obtain discounted insurance for property known to be in harm's way. The climate experts claim that reimbursement for property loss wrongly encourages victims of disastrous weather to build again in the wrong place. "Federal disaster policies, while providing obvious humanitarian benefits, also serve to promote risky behaviour in the long run." The climate scientists also stressed that storms like Katrina "are inevitable even in a stable climate." One of the scientists, Philip J. Klotzback, a hurricane researcher at Colourado State University who does not support the notion that global warming is spurring stronger or more frequent hurricanes, stated, "The social and economic trends are completely clear.

There is likely to be destructiveness in tropical storms anyway because more people now live in vulnerable coastal areas." Kerry A. Emanuel, a climatologist at the Massachusetts Institute of Technology who does believe that the building energy of hurricanes in recent decades is related to human-driven warming of the seas, also warns people from building in these areas.

Index

A
Academy 142
Administration 201
Agriculture 155
Although 12
Americans 148
Amount 159
Analyzing 160
Antarctica 72
Anthropogenic 24, 80
Anticorrelation 48
Applications 38
Approach 121
Approaches 68
Approximately 60, 111
Aquaculture 198, 201
Assessed 51
Assessment 109
Associated 24, 114, 174
Association 25
Atlantic 141
Atmosphere 73
Atmosphere 59, 68, 136
Atmospheric 27, 65, 90
Attributed 132
Availability 14, 160

B
Background 122
Backward 128
Because 131
Beginning 100, 111
Behaviour 58
Behaviours 79
Benefited 5
Between 7, 40
Biodegradable 163
Biogeochemical 38
Blossomed 20
Boundaries 173
Breach 98
Bromine 68

C
Calculation 13
Canada 33
Carbonaceous 95
Catastrophe 21
Change 106
Changing 167
Chemical 78
Chemicals 150
Chemistry 70
Chloroform 58, 67
Circulation 54, 75, 92, 155, 199
Cleaning 36
Combination 103
Combine 140
Common 173
Commonly 123

Comprehensive 90
Concentration 67, 71, 128
Concentrations 54, 64, 65, 87, 195

D

Damaged 134
Dansgaard 112
Deficiency 121
Dependent 158
Depletion 70, 83, 151
Destructive 69
Destructiveness 203
Detailed 182
Determine 122, 169
Determining 87
Developing 8
Development 84, 85
Developments 137
Difference 178
Differences 60
Discrepancies 77
Discussed 177, 181, 183
Disintegration 157
Displayed 90
Distribution 202
Documented 82
Dominates 133
Doubling 55
During 68, 146
Dynamical 77, 84

E

Ecosystem 172
Education 189
Efficacies 185
Emissions 72
Employed 83
Encroaching 70
Enhanced 46
Enhancement 45
Enough 88
Environment 138, 166

Environment 125, 200
Environmental
  18, 32, 86, 119, 175, 176, 179
Environmentalists 147
Environmentally 197
Equilibrium 131, 180
Eruptions 71
Especially 179
Established 184
Estuarine 200
Evaporation 155
Evidence 183
Examined 180
Example 11
Expected 132, 149
Experienced 6, 106
Extrinsic 188

F

Fertilizers 141
Fingerprint 22
Fourfold 101
Fragments 127
Frequency 159
Freshwater 102
Function 123, 187
Furthermore 63

G

Generally 80
Generated 139
Government 9, 197
Greater 81, 152
Greatest 149
Greenhouse 1, 19, 21, 23, 70
Greenland 108
Groundlevel 37
Growing 4

H

Halocarbons 66
Halogens 43

Index

Heating 52
Hemisphere 34
Hemispheres 157
Hierarchy 73
Higher 157, 168
Highlighted 96
Himalayan 161
Himalayas 126
Homogeneous 152
Homogenisation 116
However 110
Humidity 123
Hypothesis 141

I

Identifying 196
Immediate 92
Importance 91
Importantly 29
Impossible 146
Including 159
Incoming 39
Incorporate 103
Increase 9, 47, 148
Increased 112
Increases 14, 65, 175
Information 23, 113, 169
Integrates 95
Intensity 88, 102
Interactions 165
Interannual 49, 57
Intergovernmental 42
International 41
Interwoven 177
Inundate 129

L

Latitudinal 62
Leading 45
Leaving 168
Limatology 125
Limitations 74

Limpasuvan 56
Literally 153
Loading 76

M

Magnitude 101
Measured 115
Measurement 113
Measurements 57, 58, 143
Measuring 156
Measurments 16
Mechanism 62, 147, 184
Metabolism 198
Meteorological 139
Microclimate 89
Midwinter 49
Milankovitch 88
Misinformed 171
Mitigating 98
Modelling 130
Monitored 61
Montreal 40
Mountain 93, 172
Mutually 26

N

National 192
Natural 199
Nevertheless 72
Nitrous 10
Northern 64
Northwestern 104

O

Observational 29
Observed 74
Observers 93
Oceanographic 17
Operational 17
Organisms 187
Oxygen 37

P

Palaeoclimate 92
Palaeoclimatic 91
Parameteorological 119
Parametrizations 182
Particularly 80
Penguins 99
Peninsula 127
Periodicities 103
Persistence 117
Persistent 3
Perspective 117
Perturbation 181
Photochemical 147
Photochemistry 75
Physical 118
Phytoplankton 153
Pollutants 82
Pollution 13
Possibly 6
Precipitation 23, 25, 114, 156
Prediction 12

R

Radiation 31, 34, 150
Randomness 103
Reaction 46
Recommend 15
Reconstruct 91, 125
Reconstructions 92
Recovery 144
Regional 52
Relatively 47, 128
Reproduction 198
Research 196
Resources 1
Respectively 186
Response 20, 98, 104, 181
Responses 164
Responsible 67

S

Salinization 7
Satellites 19
Scenarios 131
Scheduled 101
Schneider 99
Seasonality 76
Species 167
Spectroradiometre 127
Statistically 139
Stratosphere 32, 35, 37, 71, 137, 151
Stratospheric 47
Stratospheric 41, 55, 61, 66, 132, 151
Strongly 87
Subsequent 31
Subtracted 19
Sufficiently 97
Suggest 186

T

Temperate 14
Temperature 199
Temperature
    26, 28, 29, 106, 115, 129, 173, 183
Temperatures 13
Temperatures 4, 15, 20, 53, 63, 168
Terminates 135
Terminations 91
Therefore 87
Traditionally 109
Transdisciplinary 176
Transformation 59
Transplant 191
Tropopause 130
Tropospheric 50, 60, 94

U

Ultraviolet 33, 36, 136
Uncertainties 98, 156
Uncertainty 48
Understanding 51, 73, 105

Index

Uniformly 145
University 105, 138
Unusually 79

V

Validation 105
Vapour 60
Variability 21, 29, 79, 85, 89, 124
Variation 93
Variations 121

Variety 179
Vegetation 158
Vulnerable 8

W

Warming 10, 11, 76
Westerly 71
Wildfires 107
Wintertime 46
Worldclass 190
Would 193